D0186848

Graphene *in* Composite Materials

Synthesis, Characterization and Applications

Nikhil A. Koratkar, Ph.D.
John A. Clark and Edward T. Crossan Chair
Professor, Mechanical, Aerospace and Nuclear Engineering
and Materials Science and Engineering
Rensselaer Polytechnic Institute

Graphene in Composite Materials

DEStech Publications, Inc.
439 North Duke Street
Lancaster, Pennsylvania 17602 U.S.A.

Printed in the United States of America
10 9 8 7 6 5 4 3 2 1

Main entry under title:
Graphene in Composite Materials: Synthesis, Characterization and Applications

A DEStech Publications book
Bibliography: p.
Includes index p. 187

Library of Congress Control Number: 2013935684
ISBN No. 978-1-60595-056-3

Table of Contents

Preface

SINCE its rise to prominence in 2004, graphene has captivated the attention of the worldwide scientific community, including academia, government, and industry. Perhaps no other material has had such an impact in such a short period of time. The remarkable blend of mechanical, electrical, thermal, optical, and wetting properties of graphene stem from its unique two-dimensional single-carbon-atom-thick hexagonal honeycomb lattice. Graphene is the thinnest, stiffest, and strongest material known to humankind and is also an excellent conductor of both heat and electricity. It is an ideal impermeable membrane that can passivate surfaces onto which it is coated. The basal plane of graphene is also extremely inert and highly resistant to oxidation. In many of these qualities, graphene bears a resemblance to carbon nanotubes. However, there are two important differences. First, graphene is two-dimensional, so it is far easier to work with compared to one-dimensional nanotubes and nanowires. It is much simpler to manipulate graphene on surfaces and to contact them electrically for various device applications. Second, carbon nanotubes do not exist in nature, while graphene is available in bulk graphite. This means that, while carbon nanotubes have to be assembled atom-by-atom using tedious, time-consuming, and expensive bottom-up synthesis techniques, graphene can be produced in bulk using relatively inexpensive and scalable top-down synthesis methods, such as exfoliation of graphite oxide or other graphite intercalation compounds. This potentially gives graphene a tremendous advantage over other forms of nanomaterials in terms of its cost-effectiveness for commercial applications.

The 2010 Nobel Prize in Physics awarded to Andre Geim and Konstantin Novoselov of the University of Manchester, UK, has directed great attention towards graphene and its potential applications. In my view, the most promising and technically feasible application of graphene for near-term practical implementation is as a nanofiller in composites. This includes polymer, ceramic, as well as metal matrix composite materials. Such materials are the basic building blocks of most engineering systems and devices. Graphene shows enormous potential to improve the mechanical, thermal, and electrical properties of such composites at ultra-low nanofiller loading fractions. This, coupled with the cost-effectiveness of graphene production (i.e., via top-down exfoliation of graphite), could result in the next generation of composite materials. These materials could find a myriad of applications, from lightweight aerospace structures used in aircraft, rotorcraft, and spacecraft, to adhesives, paints, thermal interface materials, wear- and scratch-resistant coatings, and construction materials for automobiles, boats, and building applications. The sky is the limit for such advanced composite materials and the possible uses of such technologies are only limited by our imagination.

This book is aimed at introducing graphene composites to engineering graduate and undergraduate students and academics, as well as industry and government researchers. The book is essentially a compilation of over a dozen research papers published by my group at the Rensselaer Polytechnic Institute over the last several years on this topic. The essential elements of these papers, along with other relevant work performed by other research groups, have been distilled and organized in this book into five chapters. I have attempted to simplify the language and explanation of the materials so readers who may possess only a rudimentary knowledge of graphene and its composites can easily comprehend it. The first chapter introduces the reader to the history of graphene, its place among the family of carbon allotropes, its properties, its synthesis, and its characterization methods. Chapter 2 is the main focus of this book and describes how graphene platelets and graphene nanoribbons can be infiltrated into bulk polymer systems to improve their mechanical, thermal, and electrical properties. The mechanical properties considered include Young's Modulus, ultimate tensile strength, viscoelastic properties, bucking resistance, creep response, and wear resistance, as well as basic fracture mechanics properties such as fracture toughness and fatigue resistance. The thermal conductivity and electrical conductivity enhancements induced by formation of graphene percolation networks in polymers are also considered in Chapter 2. In many realistic applications, microfibers are necessary to carry the load and, hence, Chapter 3 discusses how graphene-infused epoxy resins can be paired with conventional carbon, glass, or Kevlar fibers to create unique hierarchically organized composites. In particular, the fatigue life properties of such hierarchical materials show dramatic improvements when compared with traditional fiber-reinforced composites. In Chapter 4, I discuss the possibilities for extending the application of graphene beyond polymer systems to include ceramic, as well as metal-matrix composites. Chapter 5 describes an unconventional type of graphene composite–namely nanofluids with a fluid serving as the liquid matrix. Such graphene colloidal dispersions can be stable for extended periods of time and could find a wide range of possible applications, from cutting fluids and coolants, to coatings with controllable wetting properties, to nanofuels for enhanced combustion.

My intention is to provide the reader with a broad perspective of the possibilities and the limitations of graphene-based composite materials. The field is still young and is rapidly evolving. Thousands of researchers worldwide are engaged in graphene composites research and are constantly pushing the boundaries of what is known regarding graphene and its composites. My hope is that this book will benefit the graphene research community and industry. This book would not have been possible without the efforts of my dedicated Ph.D. students (both past and present). I have been truly blessed to have had the opportunity to work with such an energetic, ambitious, and talented group of students. In particular, I would like to thank Mohammad Ali Rafiee, Fazel Yavari, Iti Srivastava, Ardavan Zandiatashbar, Javad Rafiee, Prashant Dhiman, Abhay Thomas, Rahul Mukherjee, Ajay Krishnamurthy, Eklavya Singh, and Jing Zhong. Special thanks also to my faculty collaborators including Professor Zhong-Zhen Yu (Beijing Institute of Chemical Technology, China), Professor Pulickel Ajayan and Professor James Tour (Rice University, USA), Dr. Stephen Bartolucci (US Army Benet Labs, USA), Professor Erica Corral (University of Arizona, USA), and my colleagues Professor Theodorian Borca Tasciuc, Professor Catalin Picu, Professor Linda Schadler, and Professor Yunfeng Shi (at Rensselaer Polytechnic Institute, USA). I am deeply indebted to these individuals for many stimulating discussions and for sharing their knowledge, wisdom, and insight with me. Finally, I would like to express my deep love and gratitude to my family—specifically, my mother (Nirmala), father (Ashok), wife (Rashmi), and children (Mihir and Savani). Your patience, support, encouragement, and sacrifices have enabled me to pursue my dreams in an unfettered manner and I dedicate this book to all of you.

CHAPTER 1

Introduction to Graphene

BEFORE delving into the topic of graphene-based composite materials, it is essential to understand the structure/properties of graphene and graphene's place in the family of nanocarbon materials. The intent of this chapter is to provide the reader with a general introduction to the various allotropes of carbon that range from the well-known diamond and graphite, to newly discovered nanocarbons such as fullerenes, single-walled carbon nanotubes, multi-walled carbon nanotubes, and graphene. Graphene is, in fact, the basic building block of all forms of sp^2 hybridized carbon materials and is, therefore, of great interest both from the scientific and technological standpoint. This chapter discusses the structure of graphene and some of its key mechanical, electrical, thermal, and optical properties. It addresses the synthesis of graphene, considering both top-down and bottom-up methods for its production. This is followed by graphene characterization methods, including both microscopy- and spectroscopy-based techniques. Finally, this chapter covers why graphene is particularly promising as a nanofiller in composite materials. This lays the foundation for the subsequent chapters of this book, which cover various aspects of graphene-based composite materials.

The bulk of the material included in this chapter has been adapted from References [1–69], published by the author's group, his collaborators and other researchers.

1.1. ALLOTROPES OF CARBON

1.1.1. Diamond and Graphite

The electronic structure of carbon gives rise to its ability to bond in many different configurations and form structures with distinctly different characteristics. This is clearly manifested in diamond and graphite [1], which are the two most commonly observed forms of carbon. Diamond forms when the four valence electrons in a carbon atom are sp3 hybridized (i.e., all bonds shared equally to four neighboring atoms), which results in a three-dimensional (3D) diamond cubic structure. Diamond is the hardest material known to humankind due to this 3D network of carbon-carbon (C–C) bonds. It is also special in that it is one of the very few materials in nature that is both electrically insulating and thermally conductive. On the other hand, graphite is the sp^2 hybridized form of carbon and contains only three bonds per carbon atom. The fourth valence electron is in a delocalized state, and is consequently free to float or drift among the atoms, since it is not bound to one particular atom in the structure. This creates a planar hexagonal structure (called graphene) and gives rise to the layered structure of graphite that is composed of stacked two-dimensional (2D) graphene sheets. Graphite contains strong covalent bonds between the carbon atoms within individual graphene sheets, which gives rise to its outstanding in-plane mechanical properties. However, the van der Waal's forces between adjacent graphene sheets in the layered structure are relatively weak and, therefore, graphite is much softer than diamond. Similar to diamond, graphite (in-plane) is a good conductor of heat; however, the free electrons present in graphite also endow it with high in-plane electrical conductivity, unlike diamond. The structure of diamond and graphite are depicted schematically in Figure 1.1.

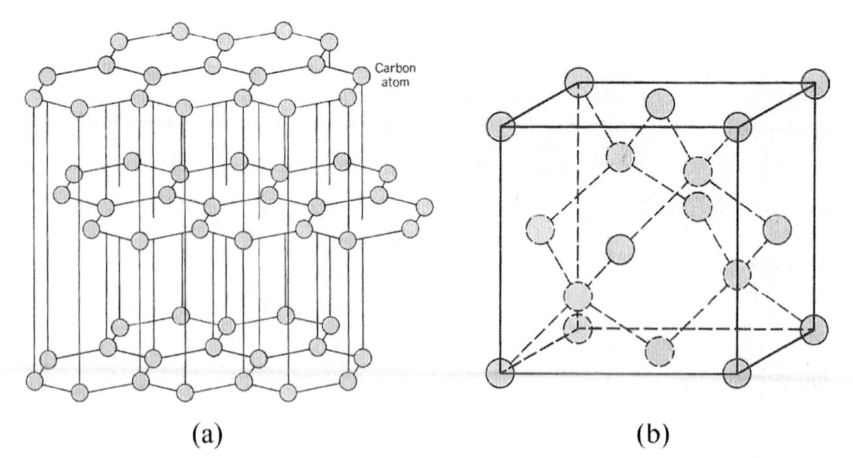

FIGURE 1.1. *Schematic of the atomic structure for (a) graphite, showing the sp^2 hybridized hexagonal lattice, and (b) sp^3 hybridized diamond, which consists of the 3-dimensional diamond cubic lattice. (Adapted from [1] with permission).*

1.1.2. Fullerenes and Carbon Nanotubes

In the last three decades, various exotic forms of nanocarbon materials have been discovered and this has revolutionized carbon science. Before the advent of graphene, fullerenes and carbon nanotubes were the most well-known nanocarbon allotropes. Both of these materials utilize the sp^2 hybridization of carbon to create self-contained molecules containing several tens or hundreds of carbon atoms. Richard Smalley and Harold Kroto first discovered fullerenes [2] in the mid-1980s. The most common of these molecules, C_{60}, has the structure of a soccer ball, containing 20 hexagon and 12 pentagon faces, with carbon atoms at each vertex. This structure is shown in Figure 1.2(a) and compared to amorphous carbon [Figure 1.2(b)], which is yet another carbon allotrope. It has been shown that pentagonal defects create a curvature in the 2D graphene structure and, with six pentagons, a complete hemisphere of a fullerene is created. Thus, 12 pentagons are necessary to form a fullerene structure. Such closed nanocarbon structures appear to be stable due to the high energy of dangling bonds at the edges of nano-sized graphene sheets. Because the free energy decrease due to satisfying these bonds is larger than the increase in bond energy by distorting the C–C bond length and angles, enclosed graphitic carbon structures such as fullerenes are created.

It can be conceived that by adding atoms to the basic structure of a fullerene, larger spherical or oblong structures are possible. In 1991, Sumio Iijima [3] discovered the carbon nanotube (CNT), the most prominent of these novel carbon materials. A single-walled carbon nanotube (SWNT) can be simply considered as a graphene sheet that has been rolled up seamlessly into a tube, and capped at the end with six pentagonal defects, as in fullerenes. By changing the way the graphene sheet is rolled into a tube [3–6], SWNT can be formed with different diameters and chiralities (or helicities). Each nanotube is characterized by a helical angle, or a chiral vector, which represents this direction of rolling. Two unique SWNT types are the zig-zag tube, which is characterized by a chiral vector of $(n,0)$ and has C–C bonds oriented perpendicular to the tube axis; and an armchair tube, which has a chiral vector of (n,n), and C–C bonds parallel to the tube axis. Figure 1.3(a)–(c) provide schematic examples of an armchair, a zig-zag, and a randomly oriented SWNT. Multi-walled carbon nanotubes (MWNTs) consist of a number of concentric SWNT cylinders, which share a common axis. The inter-shell spacing in MWNTs is ~3.4 Å, slightly larger than

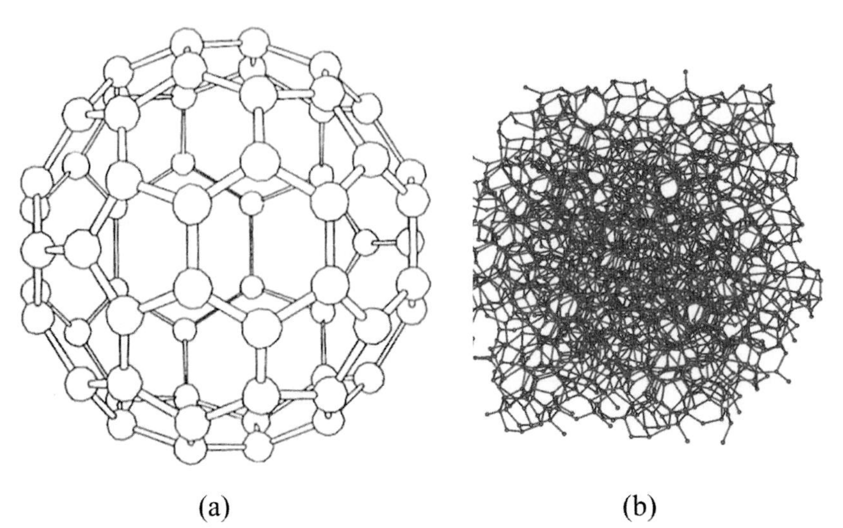

FIGURE 1.2. *(a) Soccer ball-like shape of a C_{60} fullerene molecule showing the hexagonal graphite-like lattice with pentagon defects closing the spherical shell structure. (b) Molecular model of an amorphous carbon cluster. (Adapted from [4] with permission).*

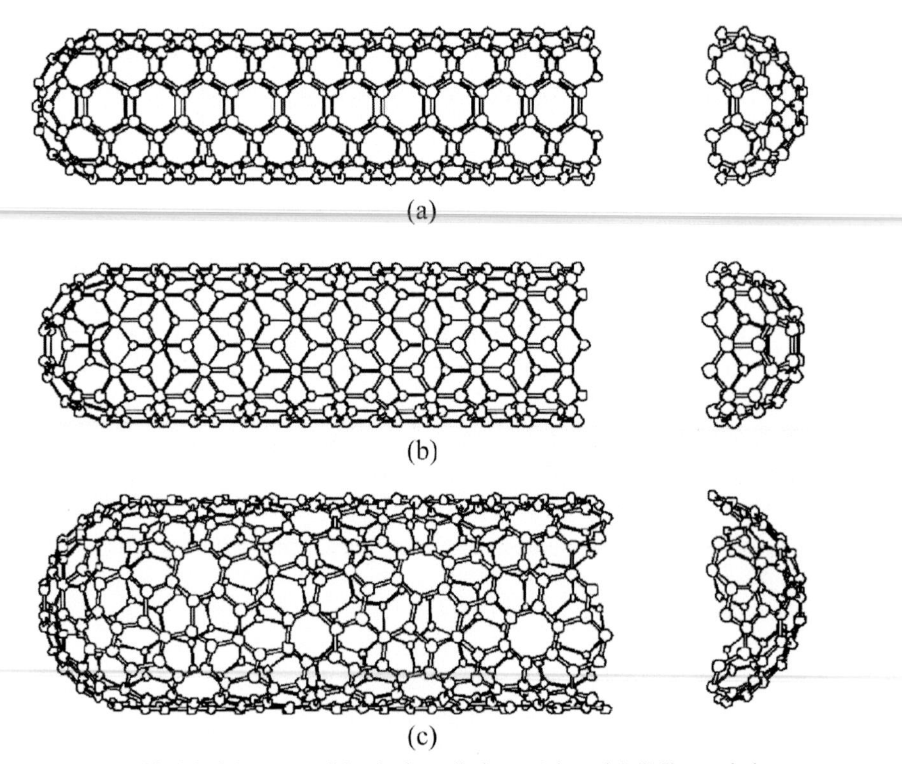

FIGURE 1.3. *Model of three possible single-walled nanotubes; (a) (5,5) armchair nanotube, (b) (9,0) zig-zag nanotube, and (c) randomly oriented (10,5) nanotube. (Adapted from [6] with permission).*

that of graphite, which has an ideal inter-plane spacing of ~3.35 Å [6]. Typical SWNT diameters lie in the 1–2 nanometer (nm) range, while MWNTs are relatively larger with diameters in the 20–40 nm range. The lengths of SWNT and MWNT can be as large as several hundred microns, thus carbon nanotubes constitute ideal one-dimensional (1D) structures. SWNT can be either semi-conducting or metallic, depending on their chiral vector, while MWNT are typically always metallic in nature. Fullerenes and carbon nanotubes can be produced by laser ablation of a carbon source. Carbon nanotubes can also be produced by arc-discharge and chemical vapor deposition processes.

1.1.3. Graphene

Graphene is a single-atom-thick sheet of sp^2 hybridized carbon atoms, which are packed in a hexagonal honeycomb crystalline structure [7–14]. Graphene is the thinnest material known to humankind; the atomic diameter of a carbon atom is ~0.14 nm. The in-plane dimensions of a graphene sheet can be several cm in size and, hence, it represents the ideal 2D sheet structure. Graphene is the fundamental building block of all sp^2 carbon materials including SWNTs, MWNTs, and graphite and is, therefore, interesting from a fundamental standpoint, as well as for practical applications. It was long believed that while graphene can be deposited on substrates (for example, using epi-

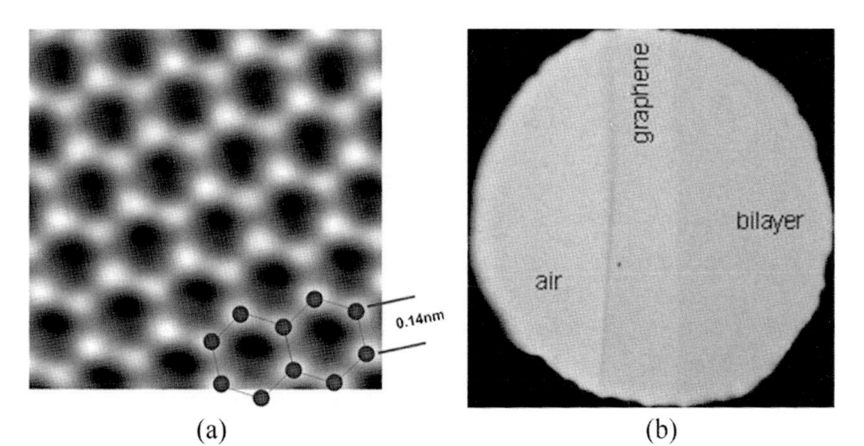

FIGURE 1.4. (a) Transmission electron microscopy image of graphene illustrating the crystalline lattice (bond length ~0.14 nm). (Adapted from [15] with permission). (b) On Si substrates with ~300 nm oxide layer, graphene layers can be discerned using an optical microscope. (Adapted from [16] with permission).

taxial growth techniques), a free-standing (non-supported) graphene sheet would be thermodynamically unstable due to the high energy of dangling bonds at the edges of the sheet. Pioneering research in 2004 by Geim and Novoselov (2010 Nobel Prize winners in Physics) of the University of Manchester led to the isolation of a single free-standing sheet of graphene [7]. Geim and Novoselov used Scotch tape to peel layers of a bulk graphite crystal. They repeated this process until they were able to isolate a graphene monolayer (Figure 1.4).

1.2. PROPERTIES OF GRAPHENE

As a result of its unique two-dimensional crystal structure and ultra-strong sp^2 carbon bonding network, graphene offers an exciting blend of mechanical, electrical, thermal, and optical properties that opens the door to a variety of possible applications. The elastic modulus of an individual graphene sheet is predicted [17] to be ~1 TPa (or 1000 GPa). This has been validated [18] by atomic force microscopy (AFM)-based indentation experiments performed on suspended graphene. The exceptionally high modulus of graphene, coupled with its low density (~1–2 g/cm^3), implies that the specific modulus (i.e., modulus normalized by density) of graphene far exceeds that of all other structural materials, including aluminum, titanium, or steel. In addition to its very large elastic modulus, graphene also displays a fracture strength of ~125 GPa [18], which is superior to most commonly used structural materials.

Graphene has a very interesting electronic band structure. It is a semi-metal with zero electronic band gap; the local density of states at the Fermi level is zero and conduction can only occur by the thermal excitation of electrons [9]. However, an energy gap can be engineered in graphene's band structure using a variety of methods. These methods are based on the breaking of graphene's lattice symmetry, such as defect generation [19], water adsorption [20], applied bias [21–23], and interaction with gases (e.g., nitrogen dioxide or ammonia) [24]. Other remarkable properties of graphene that have been reported [25] include exceptionally high values of its in-plane thermal conductivity (~5,000 W $m^{-1}K^{-1}$), charge carrier mobility (200,000 $cm^2\ V^{-1}\ s^{-1}$), and specific surface area (2,630 $m^2\ g^{-1}$), plus fascinating phenomena such as the quantum Hall effect, spin resolved quantum interference, ballistic electron transport, and bipolar super-current, to name a few. It should, however, be noted that the exceptional thermal and electronic properties of graphene listed above typically hold only for free-standing or suspended graphene and degrade markedly when the graphene is supported on a substrate platform. The number of journal publications related to graphene has increased exponentially over the past decade and there is no doubt that graphene has captivated the attention of the worldwide scientific community.

Another key property of graphene that is particularly important for optoelectronic applications is its optical transparency [26–28]. The optical absorption of a single graphene layer is shown to be ~2.3% over the visible spectrum, which, combined with its high electrical conductivity [29–30], could lead to transparent conductive electrodes [31–34]. Such electrodes could prove to be a viable replacement for transparent indium tin oxide (ITO)-based electrodes. ITO is brittle and, therefore, cannot be used in flexible electronics. Graphene, on the other hand, offers extreme flexibility; for this reason, optically transparent and electrically conductive graphene films could replace ITO for the next generation of flexible and stretchable electronics. In addition to optical transparency, another fascinating property of graphene is its wetting transparency [35]. Monolayer graphene coatings do not significantly disrupt the intrinsic wetting behavior of surfaces for which surface-water interactions are dominated by van der Waals forces. Figure 1.5(a)–(b) show water contact angle measurements performed on copper substrates with different numbers of graphene layers in between. The contact angle of water is obtained by drawing a tangent to the water droplet as it meets the solid surface. For complete wetting, the contact angle is zero, since water spontaneously wets the surface to form a liquid film. By contrast, if water completely dewets the surface, it will form a ball that will make a point contact with the solid surface, resulting in a contact angle of 180 degrees. Generally, surfaces with water contact angle > 150 degrees are called super-hydrophobic and surfaces that display contact angle < 20 degrees are termed super-hydrophilic. Figure 1.5(b) indicates that monolayer graphene causes minimal disruption in the baseline copper contact angle. With increasing numbers of graphene layers, the contact angle of water on copper gradually transitions towards the bulk graphite value, which is reached for ~6 graphene layers. Similar response has also been observed on gold and silicon, but not on glass [35].

The wetting transparency of graphene films can be comprehended from continuum modeling using the effective interface potential approach [36–39]. The basic idea is to consider the solid-liquid interfacial energy $W(h)$, defined as the excess free energy per unit area it takes to bring two interfaces from infinity to a certain distance h. By this

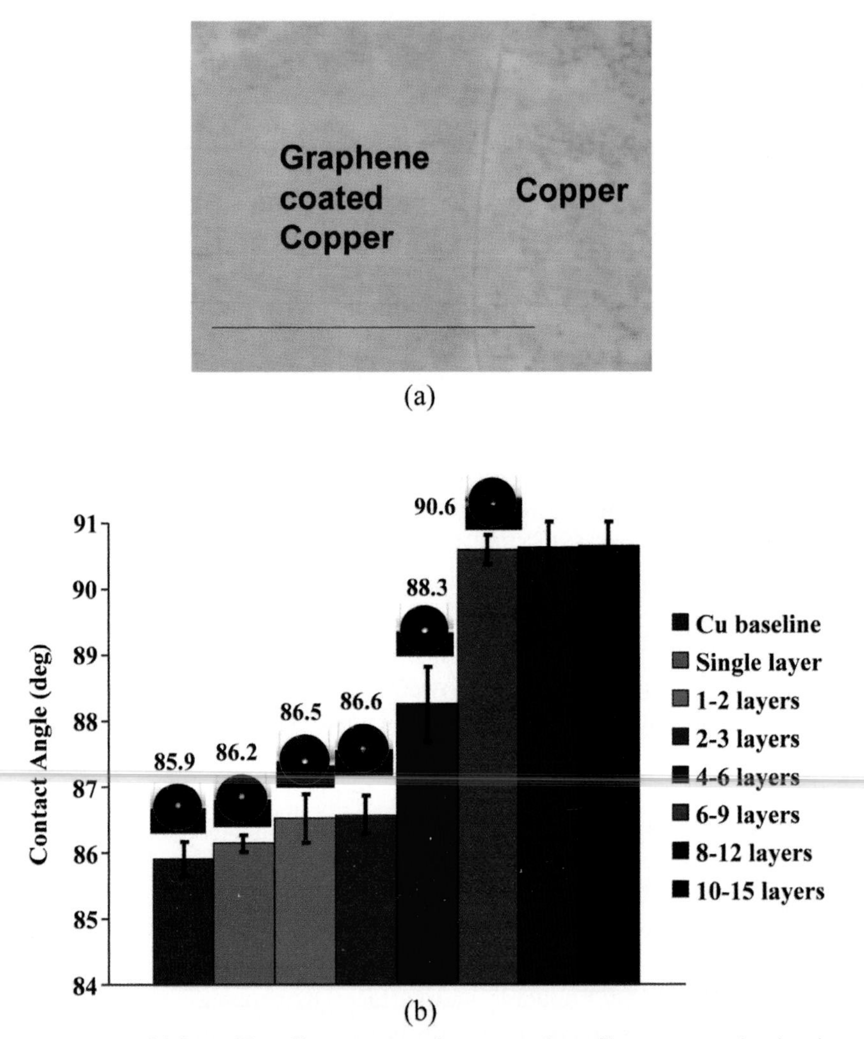

FIGURE 1.5. *(a) Deposition of large area continuous graphene films on copper by chemical vapor deposition. (b) Water contact angle measurements on copper with different number of graphene layers placed on copper. (Adapted from [35] with permission).*

definition, W can be calculated from the integration of molecular pairwise interactions across the interface. The relation between the work of adhesion (which equals the negative of interfacial energy), surface free energy, and contact angle (θ) can be described by the Young-Dupre equation [36]:

$$\gamma(1+\cos\theta) = W_{ad} = |W(h)| \tag{1}$$

The surface tension of water γ is 0.072 J/m^2. To calculate interfacial energy W, we first consider the generic half-space fluid interacting with half-space substrate (the situation of water on copper and water on graphite). The classic model of wetting assumes that the interaction of two surfaces is the summation of all the molecular pair-wise interactions across the interface. Thus, when the van der Waals interaction is chosen in a 12-6 Lennard-Jones form, the interfacial energy can be expressed as [37]:

$$W(h) = \frac{c}{h^8} - \frac{A}{12\pi h^2} \tag{2}$$

where h is the separation between the surfaces of interest. The Hamaker constant A is taken as $A_{H_2O\text{-}Cu} = 12.2 \times 10^{-20}$ and $A_{H_2O\text{-}graphite} = 9.08 \times 10^{-20}$ J, following the $A_{12} = (A_{11}A_{22})^{1/2}$ mixing rule. Here c denotes the strength of short-range repulsion, which is taken so as to match the contact angles for water on graphite and water on copper obtained from the experiments: $c_{H_2O\text{-}Cu} = 2.52 \times 10^{-80}$ J m^6, $c_{H_2O\text{-}graphite} = 0.98 \times 10^{-80}$ J m^6. The equilibrium separations are found to be ~1.77 Å and ~1.59 Å for water on copper and water on graphite, respectively.

Next, let us consider the situation of water on the copper substrate with graphene in between. The interfacial energy for this case can be expressed as [35]:

$$W(h) = \begin{pmatrix} \dfrac{c_{H_2O-graphite}}{h^8} - \dfrac{c_{H_2O-graphite}}{(h+d)^8} + \dfrac{c_{H_2O-Cu}}{(h+d)^8} - \\ \dfrac{A_{H_2O-graphite}}{12\pi h^2} - \dfrac{A_{H_2O-graphite}}{12\pi (h+d)^2} + \dfrac{A_{H_2O-Cu}}{12\pi (h+d)^2} \end{pmatrix} \tag{3}$$

where h is the separation between water and the substrate and d is the thickness of the graphene film. When $d = 0$, Equation (3) reduces to the correct form for work of adhesion for water on Cu substrate. When d approaches infinity, Equation (3) reduces to the correct form for water on graphite. Figure 1.6 shows the prediction using Equation (3) for the water contact angle transition from bulk copper to bulk graphite with increasing numbers of graphene layers on the copper slab. The thickness of one graphene layer is assumed to be ~0.34 nm in the simulations. The wetting angle transition predicted by Equation (3) is consistent with experimental observations. This indicates that the wetting transparency of graphene is attributable to its extreme thinness. The van der Waals interaction [Equation (3)] is calculated by integrating the interaction of all the molecules in the fluid with all the molecules in the substrate, which, therefore, results in wetting transparency of the ultrathin graphene monolayer to the relatively long-range van der Waals interactions. However, it should be noted that the wetting transparency of graphene breaks down for surfaces such as glass. In spite of its extreme thinness, the presence of graphene at the water-glass interface disrupts the short-range chemical interactions (hydrogen bonding networks), which dictate the water/glass contact angle. Therefore, for surfaces where chemistry plays the dominant role, graphene coatings do not provide wetting transparency.

Due to its extreme thinness, graphene has an unparalleled ability to provide transparency to van der Waals interactions. This is illustrated in Figure 1.6, where the water contact angle transition from copper to graphite is shown for carbon film coatings on copper with thicknesses

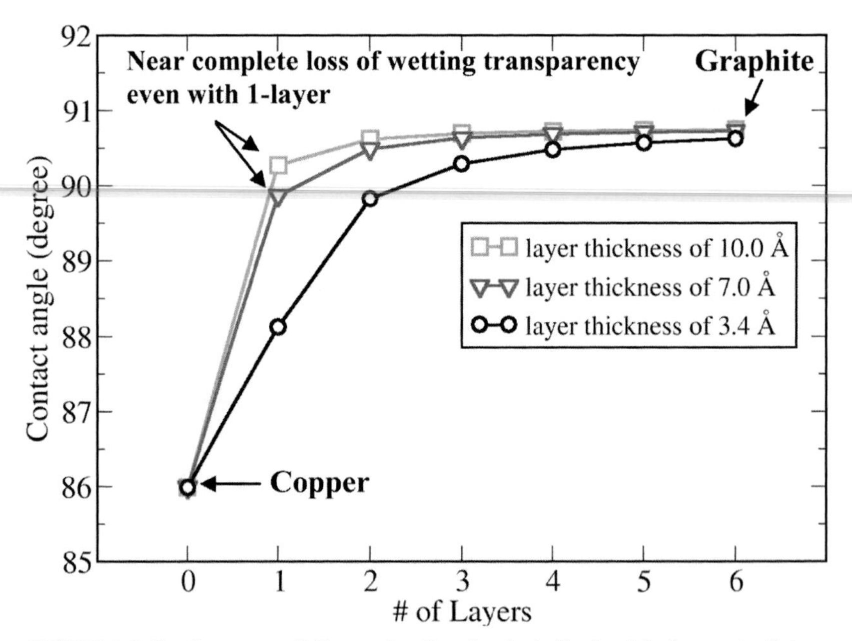

FIGURE 1.6. Continuum predictions using the classical effective interface potential approach for water contact angle transition from copper to graphite with carbon coating layers of thickness 0.34 nm, 0.7 nm, and 1 nm. As the number of layers is increased, the water contact angle increases from copper and saturates at the bulk graphite value. However, even a single 0.7 nm thick layer fails to show significant wetting transparency effect. This highlights the need for the extreme thinness of graphene (~0.3 nm) in terms of achieving wetting angle transparency. (Adapted from [35] with permission).

of 0.34 nm, 0.7 nm, and 1 nm using Equation (3). It is clear that even ultrathin 0.7 nm or 1 nm coatings fail to provide wetting transparency. The wetting transparency effect becomes apparent only when the coating thickness is reduced to ~0.34 nm (i.e., thickness of graphene). Sputtered oxide or polymer films cannot rival such extreme levels of thinness. Hexagonal boron nitride (h-BN) is the only known material system that could match graphene's thinness; however, it is very challenging to deposit monolayer h-BN on large area substrates. By contrast, roll-to-roll deposition methods have already been developed [33] to deposit monolayer graphene films several tens of inches in dimensions for flexible electronics.

Graphene also exhibits an extraordinary ability to passivate a surface. Graphene is the ideal impermeable membrane and not even a proton can pass through defect-free graphene. Therefore, conformal coating of graphene on copper prevents the copper from oxidizing [40]. This can have wide-ranging impact in heat-transfer applications, where copper is the material of choice due to its very high thermal conductivity. Copper oxide acts as an interfacial thermal barrier, which significantly degrades heat transfer across copper interfaces. This problem could be overcome by simply coating a monolayer graphene film onto copper.

1.3. SYNTHESIS OF GRAPHENE

There are four basic methods [25] used for graphene synthesis: (1) chemical vapor deposition; (2) epitaxial growth of graphene on electrically insulating substrates; (3) mechanical exfoliation of graphene from bulk graphite (e.g., using Scotch tape); and (4) reduction of graphene derivatives such as graphene oxide. Among these methods (1) and (2) can be broadly classified as bottom-up methods, while methods (3) and (4) are top-down approaches. Among these, reduction of graphene oxide and chemical vapor deposition are the two methods that show the greatest promise for bulk production of graphenes at the scale necessary for composites applications. These two methods are described in brief below.

1.3.1. Reduction of Graphene Oxide

In this approach the starting material used is graphite, which is first converted to graphite oxide using a modified Hummers method [25]. The typical procedure for this is as follows: a reaction flask with a mechanical stirrer is charged with sulfuric acid (87.5 ml) and nitric acid (45 ml) and cooled by immersion in an ice-water bath. Fifteen minutes later, graphite flakes (5 g) are introduced into the flask under vigorous stirring to avoid agglomeration. Potassium chlorate (55 g) is then added into the suspension slowly. After reacting for 96 h at room temperature, the suspension is diluted with a mass of deionized water. Graphite oxide precipitate is then washed with HCl solution (10%) to eliminate sulphate ions. Barium chloride can be used to test whether the sulphate ions are eliminated. Once this is confirmed, the graphite oxide is extracted using a high-speed centrifuge and is washed with deionized water until the pH value becomes neutral. After drying in a vacuum oven at ~80°C for ~24 h, the graphite oxide is now ready for use.

It is important to check whether the graphite has been fully oxidized to graphite oxide, as this can have a strong effect on the subsequent exfoliation of graphite oxide into graphene oxide nanosheets. For this, X-ray diffraction (XRD) is a powerful tool. Figure 1.7(a) shows the XRD patterns of natural graphite and graphite oxide. Natural graphite exhibits a strong and sharp peak at 26.5°, indicating a highly ordered structure. The calculated intra-gallery spacing of graphene sheets in graphite structure is ~0.34 nm. This peak disappears after oxidation of the graphite, while a new one arises at 12.3°, corresponding to an intra-gallery spacing of ~0.72 nm, which implies complete oxidation of graphite. The increased spacing arises from the fact that a variety of oxygen moieties including hydroxyl, epoxide, carbonyl, and carboxylic functional groups are attached to the individual graphene sheets due to the oxidation process [25]. The resulting layered structure is called graphite oxide and is composed of graphene oxide sheets (i.e., graphene functionalized with hydroxyl, epoxide, carbonyl, and carboxylic groups). The attachment of the aforementioned oxygen-containing functional groups disrupts the sp^2 bonding network in graphene, creating local sp^3 islands. Graphene oxide is, therefore, electrically insulating compared to its parent graphene.

As Figure 1.7(a) indicates, the physical separation between graphene oxide nanosheets in graphite oxide is ~0.7 nm compared to ~0.34 nm in graphite. The implication of this is that the inter-sheet van der Waals interactions in graphite oxide are significantly weaker in comparison to the original graphite structure. It is, therefore, far easier to exfoliate graphite oxide then it is to exfoliate graphite directly into graphene. There are two basic approaches to exfoliating graphite oxide. The first method involves ultrasonication of graphite oxide in water. Water is able to penetrate into the inter-layer spacing in graphite oxide because of the strongly hydrophilic nature of the oxygen moieties on graphite oxide. This causes the graphite oxide to exfoliate completely in water to produce a colloidal dispersion of individual graphene oxide nanosheets in water. In the last step, these graphene oxide nanosheets are chemically reduced to graphene in solution using reducing agents such as hydra-

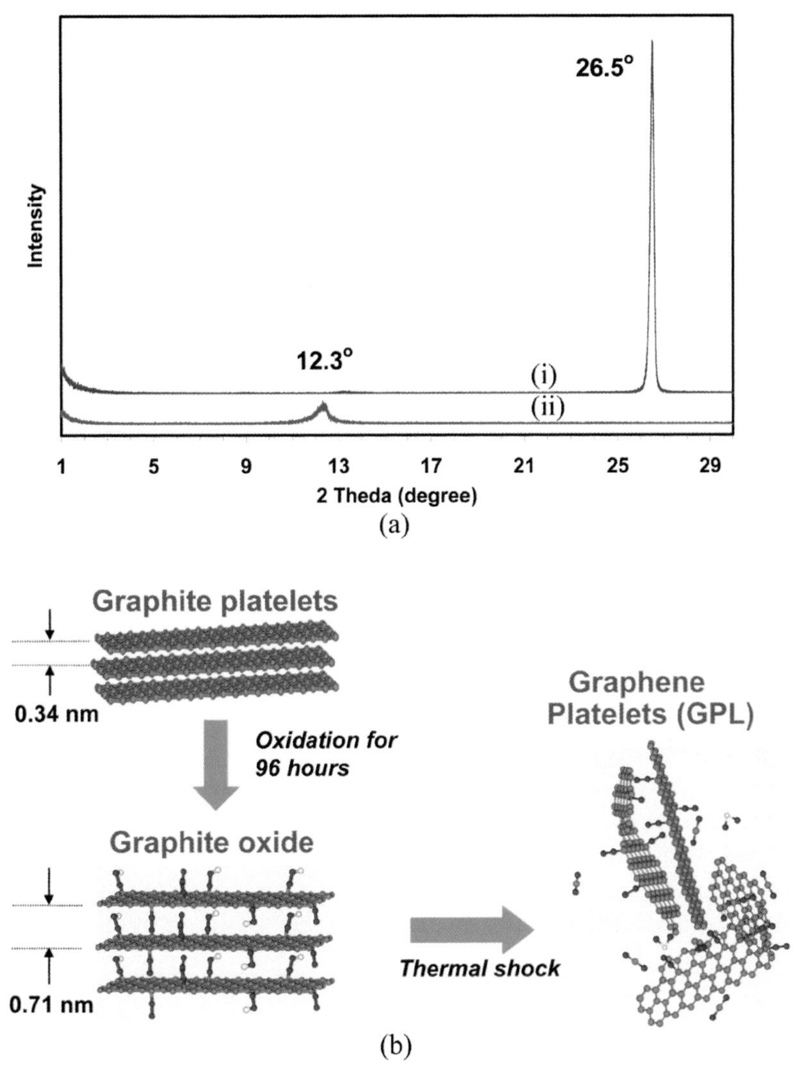

FIGURE 1.7. *(a) X-ray diffraction patterns of natural graphite (i), graphite oxide (ii). (b) Schematic representation of the method used to exfoliate graphite into bulk quantities of graphene platelets. (Adapted from [57–58], with permission).*

zine hydrate [25]. Typically, a surfactant is also needed to prevent re-stacking and agglomeration of graphenes post-reduction. The reduction of graphene oxide to graphene partially restores the sp^2 carbon bonding network, however there are residual oxygen groups still remaining on graphene. For this reason, graphene produced by this method is also called reduced graphene oxide (RGO). It is important to quantify the degree of oxygen leftover on the graphene. For this, X-ray photoelectron spectroscopy is a powerful tool. Typical carbon to oxygen ratio in graphite oxide is ~3:1, which can be increased to ~10:1 post reduction. Annealing at high temperatures in inert environments can be used to further increase the elemental carbon to oxygen ratio, although some trace amounts of oxygen will always remain no matter the treatment. In recent years, environmentally friendly approaches [41–47] for the reduction of graphene oxide, including photo-reduction, hydrothermal dehydration, and solvo-thermal reduction, as well as catalytic/photo-catalytic reductions have been developed. Recently, the substitution of hydrazine by a variety of reducing agents [48–54], including melatonin, vitamin C, sugar, and bovine serum albumin, as well as bacteria has been demonstrated.

As described above, the reduction of graphene oxide in solution involves two distinct steps: (1) the exfoliation of graphite oxide into individual graphene oxide nanosheets, achieved by ultrasonication in water, and (2) reduction of the graphene oxide to graphene using chemical, catalytic, photo-thermal, hydrothermal, or solvo-thermal methods. The thermal shock method [Figure 1.7(b)], pioneered by Aksay and co-workers [55–56] is an alternative and very elegant approach to combining these two steps and achieving simultaneous exfoliation and reduction of graphite oxide into graphene nanosheets. In this method, bulk graphite oxide powder is exposed to a thermal shock (heating rate of ~2000°C/min). The thermal shock is delivered by inserting the graphite oxide powder into a tube furnace that is pre-heated to ~1000°C and flushed with argon. The graphite oxide is kept in the furnace for ~30 seconds and then removed. Because of the rapid rate of heating, the oxygen groups released as CO_2 gas are unable to diffuse out of the layered graphite structure. This eventually results in a large pressure build up that overcomes the inter-layer van der Waals forces and exfoliates [Figure 1.7(b)] the graphite structure into individual graphene platelets (GPL). Transmission electron microscopy (TEM) characterization [57–58] of a typical GPL flake is shown in Figure 1.8(a), indicating lateral sheet dimensions of ~2–3 μm. High resolution TEM [Figure

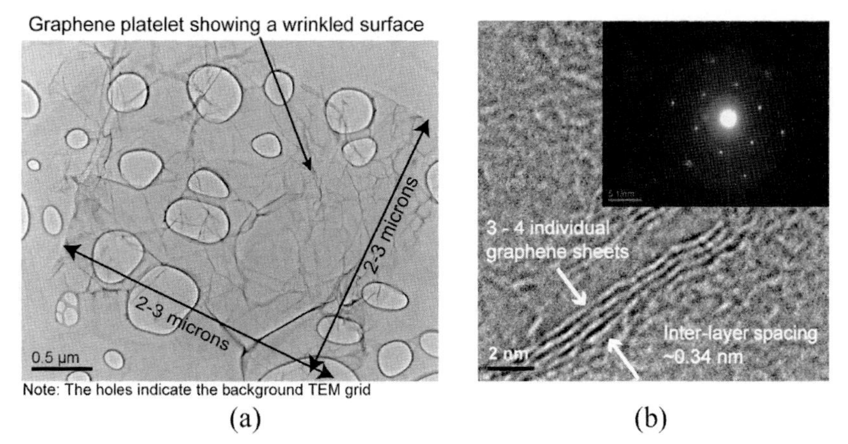

FIGURE 1.8. (a) Transmission electron micrograph of a typical GPL flake indicating lateral dimensions in the micron range. (b) High resolution transmission electron microscopy characterization of the GPL, indicating thickness of ~2 nm. The GPL appears to be composed of ~3–4 graphene sheets with an interlayer spacing of ~0.34 nm. Inset shows the measured electron diffraction pattern, confirming the hexagonal close-packed structure of graphene. (Adapted from [57–58] with permission).

1.8(b)] indicates the GPL are composed of ~3–4 individual graphene sheets within each platelet, with an interlayer spacing of ~0.34 nm. The electron diffraction pattern in the inset confirms graphene's hexagonal close-packed structure. Note that the complete exfoliation of graphite oxide into single-layer graphene sheets is very challenging. Most methods result in few-layered graphene platelets, as shown in Figure 1.8.

1.3.2. Chemical Vapor Deposition (CVD)

Graphene growth by vapor phase deposition of gaseous hydrocarbons on metal substrates such as Ni, Co, Ir, and Ru has been reported [59–62]. On Ni substrates, it is challenging to produce single-layer graphene. The reason for this is that, at the high temperatures associated with CVD, carbon dissolves into the Ni and precipitates out somewhat uncontrollably when cooled, resulting in few to multi-layer graphene. The problem of efficiently synthesizing high-quality, single-layer graphene persisted until the Ruoff group showed that, on Copper foils [63], CVD tends to produce predominantly monolayer graphene. They used methane gas as a precursor in their process. Carbon has very low solubility in copper and, hence, the mechanism of graphene growth involves adsorption of carbon atoms on the copper surface. Once a monolayer of graphene is formed, the process becomes self-limiting and further gra-

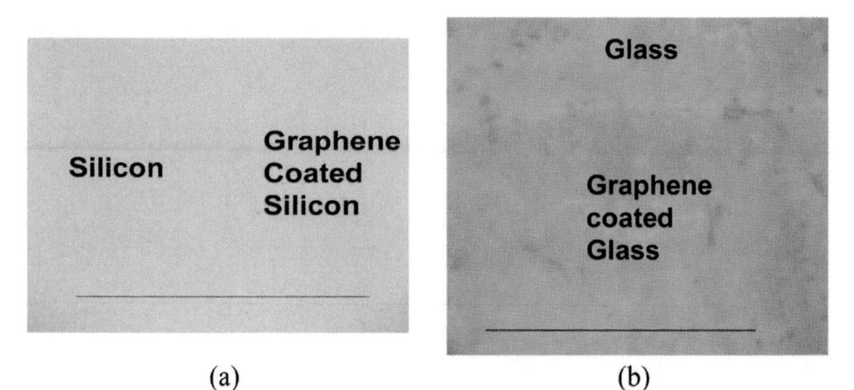

FIGURE 1.9. (a) Optical image of large area graphene film grown on copper foil and then transferred onto a Si/SiO_2 wafer. (b) Similar transfer process performed on a glass substrate. Scale bar in the images is ~1 mm. (Adapted from [35] with permission).

phene growth is suppressed. This method produces large-area, mostly single-layer graphene, but requires high vacuum to be maintained, and accurately controlling methane pressure can be challenging.

Another approach pioneered by the Ajayan group is to use liquid precursors such as hexane for CVD growth [64]. Growth using liquid precursors is advantageous because the relevant organic compounds are readily available at low cost in the liquid phase at room temperatures. In this method, Cu foils are loaded in a quartz tube and pumped down to $\sim10^{-2}$ Torr vacuum before flowing in Ar/H_2 at a pressure of ~8–9 Torr (with flow rate of ~400 sccm). Next, the sample is heated to ~950°C inside the quartz tube in Ar/H_2 ambient. At the desired temperature, the Ar/H_2 flow is stopped and hexane vapor is flowed into the quartz tube to maintain a pressure of ~500 mTorr for ~4 min. The flow rate of hexane is ~4 mL/h. The graphene films thus produced can be transferred onto a variety of substrates. For this, a thin poly-methyl-methacrylate (PMMA) film is coated onto the graphene/Cu system. The underlying copper foil is then dissolved in dilute nitric acid and the graphene/PMMA film is transferred onto the substrate of interest. After the transfer is completed, the PMMA film is dissolved away using hot acetone. Figure 1.9 shows optical microscopy images of large area graphene films deposited on Si/SiO_2 and glass substrates by using this method.

1.4. CHARACTERIZATION OF GRAPHENE

There are a variety of characterization methods that can be utilized to study the structure and properties of graphene. These techniques can be broadly classified into two categories: (1) microscopy methods and (2) spectroscopy tools. Microscopy tools are used to observe the structure of graphene at various length scales and include optical microscopy, atomic force microscopy, scanning tunneling microscopy, scanning electron microscopy, and transmission electron microscopy [65]. Spectroscopy tools are used to map out the energy/momentum distributions in graphene and include Raman spectroscopy, Auger spectroscopy, and X-ray photoelectron spectroscopy [66]. A brief description of some of these characterization methods in the context of graphene is provided below.

1.4.1. Atomic Force Microscopy (AFM)

AFM line scans are commonly used to determine the thickness of graphene sheets and also to study their surface topography. A typical AFM line scan performed along the edges of a graphene sheet is shown in Figure 1.10(a). There is a step jump in the height position of the AFM tip as it passes over the graphene sheet edge. Based on this change and the known thickness of graphene (~0.3 nm), the number of graphene sheets in the sample can be estimated. Note that the first graphene layer typically generates a ~1 nm height change; this is typically an artifact of the AFM calibration and set-up [65]. Subsequent graphene layers cause ~0.3 nm height change, as expected. AFM is also widely used to study the topography of the graphene film. Figure 1.10(b) shows a typical topography map of a graphene film grown on copper foils and then transferred onto a Si/SiO_2 wafer. The surface is highly wrinkled, with the average height of wrinkles of the order of several nm. Such wrinkles can result from the transfer process used to detach the graphene from the copper foil and transfer it to Si/SiO_2. It could also result from a mismatch in thermal expansion coefficient between the graphene film and the underlying copper foil.

1.4.2. Raman Spectroscopy

Graphene surfaces are Raman active and provide a number of characteristic Raman peaks when excited by a laser (typical wavelength used is ~514.5 nm). The G peak associated with the longitudinal vibration of the carbon atoms in graphene occurs at ~1581 cm^{-1}. The other key modes are the D band (~1310 cm^{-1}) and the 2D band (~2680 cm^{-1}), which are associated with double resonance processes [67–68]. Figure 1.11(a) shows a typical Raman spectrum for graphene platelets obtained by the exfoliation and reduction of graphite oxide. A dominant D peak is observed at ~1311 cm^{-1}, which is stronger in intensity than the G peak. Such a response with a dominant D peak is indicative of a high degree of defects and disorder in the graphene. This is expected for graphene produced from graphite oxide, since the oxidation of graphite disrupts the sp^2 bonding network. Further, when the oxygen moieties are expelled due to the thermal shock, we expect this process to be highly abrasive, leaving the structure littered with a high density of structural and topological defects such as vacancies, dangling bonds, and Stone-Wales defects, among others. Another interesting feature of the

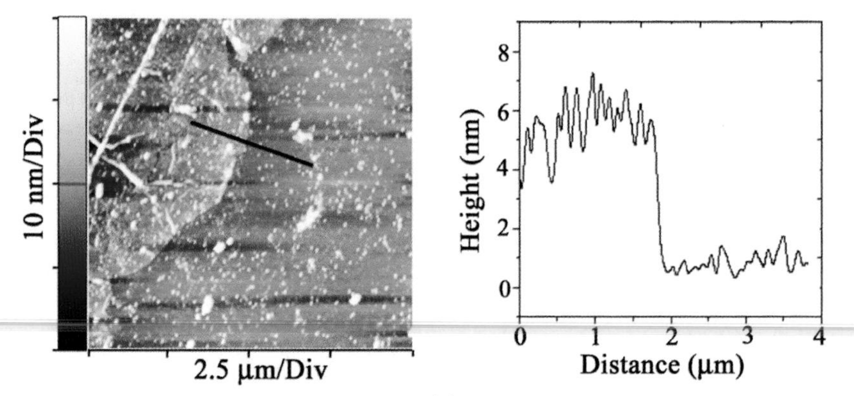

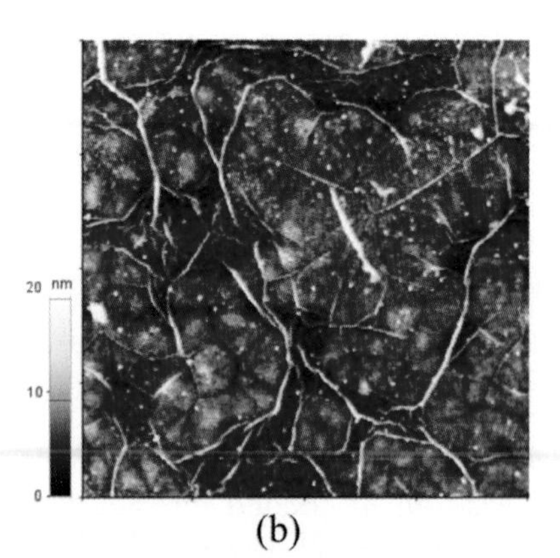

FIGURE 1.10. *(a) AFM line scan performed along the graphene sheet edge. (bottom) Topography map of graphene sheet deposited on a Si/SiO_2 substrate. (Adapted from [20] with permission).*

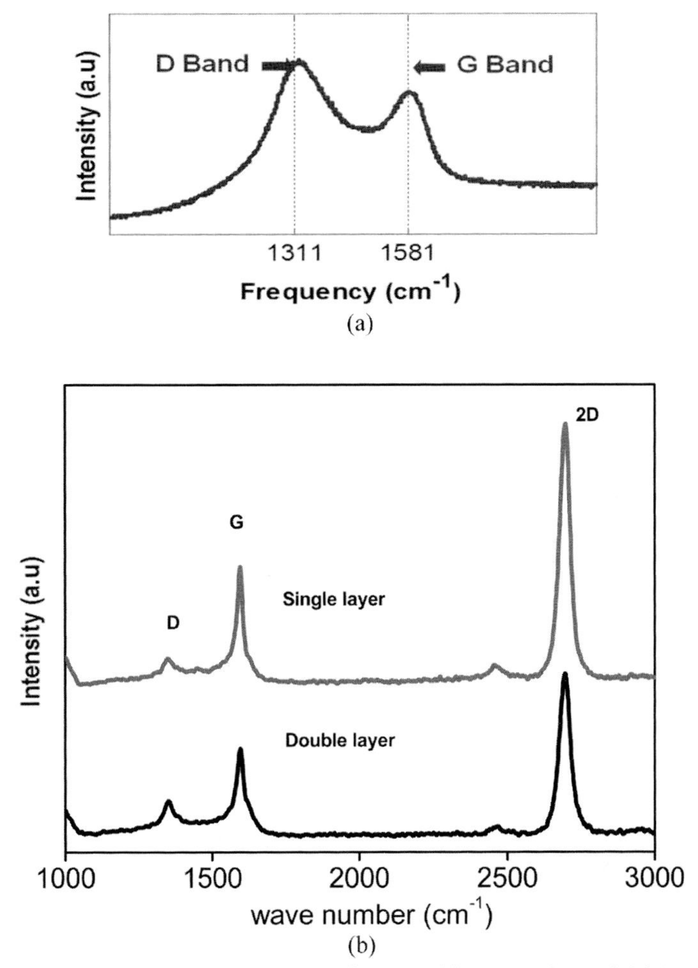

FIGURE 1.11. *(a) Raman spectra obtained from a multilayer graphene platelet produced by the exfoliation and reduction of graphite oxide. (b) Raman spectra from CVD-grown graphene on a Cu foil and transferred onto a Si/SiO*$_2$ *substrate. (Adapted from [64] with permission).*

Raman signature in Figure 1.11(a) is the significant broadening of the G and D peaks. This suggests that, in addition to defects, there are residual groups (oxygen containing moieties) that are attached to the graphene. It should be noted that the 2D peak is non-existent in Figure 1.11(a), which indicates the graphene produced by the thermal exfoliation and reduction of graphite oxide is in the form of multilayered platelets.

Figure 1.11(b) shows the characteristic Raman spectra obtained from CVD-grown graphene. The defect-related D band is strongly suppressed, which indicates high quality, defect-free graphene is produced by CVD. Further, a dominant 2D peak is observed, which is indicative of mono to bilayer graphene. For monolayer graphene the ratio of the integrated intensity of the 2D to G peak is ~3.0, while for bilayer graphene it is ~2.0. The location of the 2D peak is also known to be sensitive to the number of graphene layers in the film. Figure 1.12(a)–(b) show the position of the 2D Raman peak for graphene films with varying number of layers. A clear shift in the position of the 2D peak with number of layers is observed (Figure 1.12) from ~2680 cm^{-1} for monolayer graphene to ~2715 cm^{-1} for the >10 graphene layers.

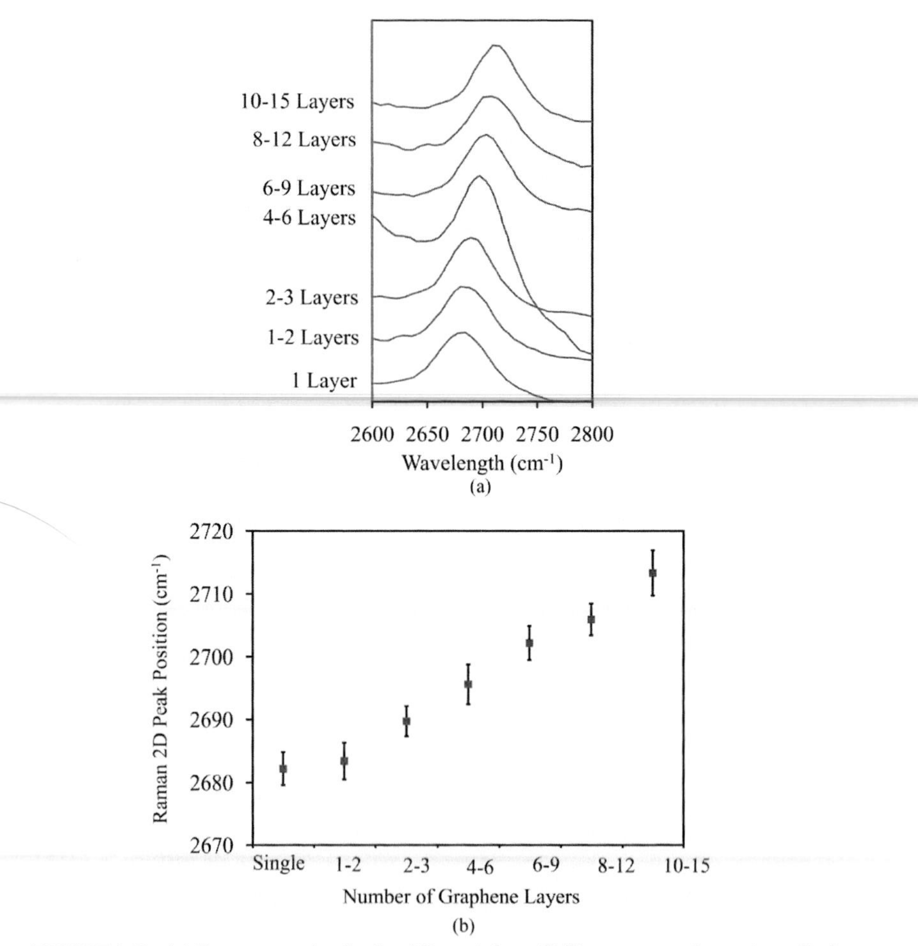

FIGURE 1.12. *(a) Raman spectra for the 2D peak from CVD-grown graphene deposited on a copper substrate. (b) The 2D Raman peak position is observed to be very sensitive to the number of graphene layers in the film. (Adapted from [35] with permission).*

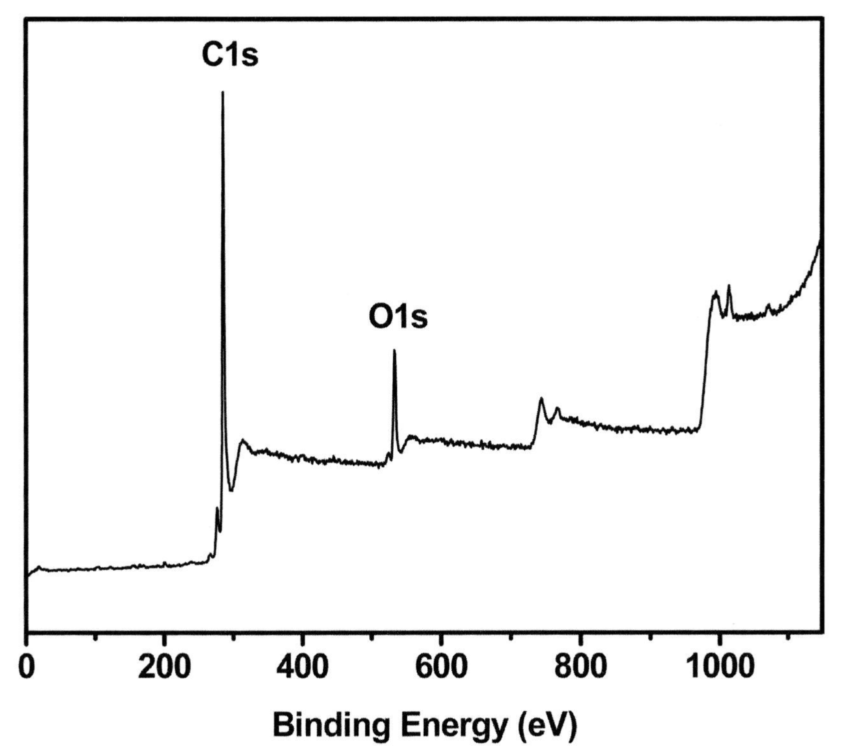

FIGURE 1.13. X-ray photoelectron spectroscopy of graphene platelets, indicating C1s and O1s peaks. (Adapted from [69] with permission).

1.4.3. X-Ray Photoelectron Spectroscopy (XPS)

XPS is widely used to study the surface chemistry states of graphene. This is particularly important for graphene grown by the exfoliation and reduction of graphite oxide [69]. Figure 1.13 shows the broad scan XPS spectra, indicating C1s and O1s peaks. The peaks in the higher energy range (around ~750 eV, ~1000 eV) are attributed to the Auguer peaks of C and O. Elemental analysis gives a carbon-to-oxygen ratio for the graphene platelet as ~9.1. This indicates that the graphene platelet is not completely reduced and contains residual oxygen. Figure 1.14 depicts the C1s XPS spectra, which is split into two distinct peaks (285 eV, 286.1 eV), and are attributed to the non-oxygenated ring C and the C in C–OH bonds, respectively. Therefore, after the thermal shock, most of the pendant oxygen moieties in graphite oxide such as carbonyl, carboxyl, and epoxide groups are eliminated, but some residual hydroxyl groups still remain.

1.5. GRAPHENE AS A NANOFILLER IN COMPOSITES

Previous sections of this chapter discussed the discovery of graphene, its place in the family of nanocarbon attotropes, and the unique structure and properties of graphene, as well as various methods to synthesize, produce, and characterize graphene. The rest of this book will focus on one specific and highly promising application of graphene–namely as a nanofiller in composites. There are a number of reasons why graphene can have a powerful effect on the mechanical, thermal, and electrical properties of the host matrix structure into which it is inserted. These reasons are summarized below:

(1) Graphene has an exceptionally high theoretical specific surface area [25] of ~2,630 m^2/g. This is because both surfaces of the graphene sheet will be in contact with the matrix, as opposed to a

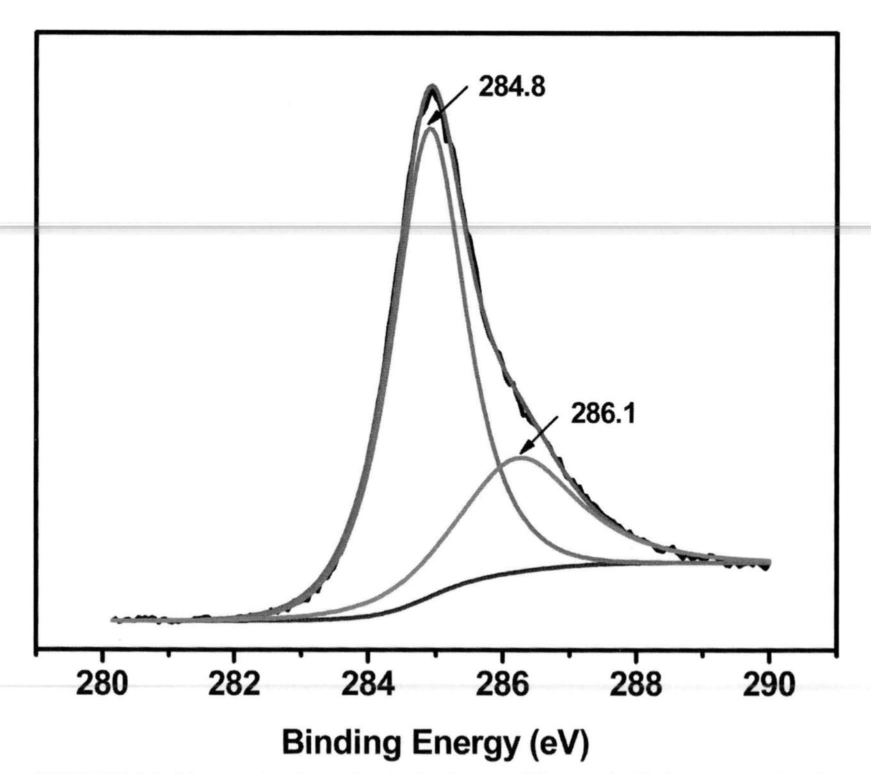

FIGURE 1.14. C1s core level spectra, indicating two fitted peaks that correspond to the non-oxygenated ring C (284.8 eV) and the C in C–OH bonds (286.1 eV). (Adapted from [69] with permission).

carbon nanotube or nanowire for which only the outer surface of the nanotube or nanowire contacts the matrix. As a consequence of this, graphene could be more effective at lower nanofiller loading fractions in the composite as compared to other competing nanofillers.

(2) The surface texture of graphene is extremely rough and wrinkled [25], which could enable mechanical interlocking with the matrix and, consequently, improved load transfer.

(3) Graphene platelets produced from graphite oxide have a high density of surface defects [65], which could provide high energy sites for the adsorption of polymer chains. As a consequence, polymers will display a stronger propensity to wet graphene surfaces.

(4) Graphene platelets synthesized from graphite oxide have residual hydroxyl groups [66], which could provide chemical handles for interaction with the surrounding matrix.

(5) The two-dimensional sheet geometry of graphene enables it to carry load equally effectively in-plane. Therefore, unlike high aspect ratio fibers, graphene platelets do not have to be perfectly aligned in the loading direction.

(6) The ultra-high aspect ratio and two-dimensional sheet geometry of graphene could significantly alter fracture and fatigue processes in polymers and ceramics due to crack defection and related mechanisms.

(7) Formation of electrical percolation networks could be formed at extremely low nanofiller loadings due to the two-dimensional geometry, sp^2 bonding structure, and large aspect ratio of graphene.

(8) The sp^2 carbon bonding network in graphene and its geometrical features of high aspect ratio and two-dimensional sheet geometry could also foster enhanced thermal conductivity compared to the more traditional one-dimensional fiber reinforcement.

(9) Graphene forms highly stable colloidal suspensions with a range of organic liquids, which creates possibilities of nanofluid-based composites that can find many exciting applications.

(10) Further, unlike carbon nanotubes, which do not exist in nature, graphene sheets are already present in graphite. Therefore, top-down methods like exfoliation of graphite oxide could be used to mass produce graphene nanosheets at low cost. Such top-down options do not exist for most other categories of nanofillers.

The balance of this book is dedicated to answering the above questions. Chapter 2 will focus on graphene-polymer composites; Chapter 3 will explore combining graphene-polymer blends with conventional microfiber reinforcement to create hierarchically organized multiscale composites; Chapter 4 will explore the possibility of developing graphene ceramic and metal-matrix composites; and Chapter 5 will explore liquid matrices for graphene, leading to novel graphene colloids and coatings. The materials included in these chapters are intended to provide the reader with a comprehensive understanding of the relative strengths and the limitations of graphene as a nanofiller in composite materials.

1.6. REFERENCES

1. Callister, W. D. Materials Science and Engineering: An Introduction; John Wiley & Sons: New York. 1997.
2. Kroto, H. W.; Heath, J. R.; O'Brien, S. C.; Curl, R. F.; Smalley, R. E. C60: Buckminsterfullerene. *Nature* **1985**, *318*, 162.
3. Iijima, S. Helical microtubules of graphitic carbon. *Nature* **1991**, *354*, 56.
4. Dresselhaus, M. S.; Dresselhaus, G.; Eklund, P. C. Science of Fullerenes and Carbon Nanotubes; Academic Press: San Diego, 1996.
5. Ajayan, P. M. Nanotubes from carbon, *Chem. Revs.* **1999**, *99*, 1787–1795.
6. Dresselhaus, M. S.; Dresselhaus, G.; Avouris, Ph. Carbon Nanotubes: Synthesis, Structure, Properties, and Applications; Springer-Verlag: Berlin, 2001.
7. Novoselov, K. S. et al. Electric field effect in atomically thin carbon films. *Science* **2004**, *306*, 666–669.
8. Novoselov, K. S. et al. Two-dimensional gas of massless Dirac fermions in graphene. *Nature* **2005**, *438*, 197–200.
9. Zhang, Y.; Tan, Y. W.; Stormer, H. L.; Kim, P. Experimental observation of the quantum Hall effect and Berry's phase in graphene. *Nature* **2005**, *438*, 201–204.
10. Li, D.; Kaner, R. B. Graphene-based materials. *Science* **2008**, *320*, 1170.
11. Geim, A. K.; Novoselov, K. S. The rise of graphene. *Nat. Mater.* **2007**, *6*, 183.
12. Schedin, F.; Geim, A. K.; Morozov, S. V.; Hill, E. W.; Blake, P.; Katsnelson, M. I.; Novoselov, K. S. Detection of individual gas molecules adsorbed on graphene. *Nat. Mater.* **2007**, *6*, 652.
13. Elias, D. C.; Nair, R. R.; Mohiuddin, T. H. G.; Morozov, S. V.; Blake, P.; Halsall, M. P.; Ferrari, A. C.; Boukhvalov, D. W.; Katsnelson, M. I.; Geim, A. K.; Novoselov, K. S. Control of graphene's properties by reversible hydrogenation: evidence for graphane. *Science* **2009**, *323*, 610.
14. Zhou, S. Y.; Gweon, G. H.; Fedorov, A. V.; First, P. N.; De Heer, W. A.; Lee, D. H.; Guinea, F.; Castro Neto, A. H.; Lanzara, A. Substrate-induced bandgap opening in epitaxial graphene. *Nat. Mater.* **2007**, *6*, 770.
15. Dato, A.; Lee, Z.; Jeon, K. J.; Erni, R.; Radmilovic, V.; Richardson, T. J.; Frenklach, M. Clean and highly ordered graphene synthesized in the gas phase. *Chem. Commun.* **2009**, *40*, 6095.
16. Kuzmenko, A. B.; van Heumen, E.; Carbone, F.; van der Marel, D. Universal optical conductance of graphite. *Phys. Rev. Lett.* **2008**, *100*, 117401.
17. Kudin, K. N.; Scuseria, G. E.; Yakobson, B. I. C2F, BN, and C nanoshell elasticity from ab initio computations. *Phys. Rev. B* **2001**, *64*, 235406.

18. Lee, C.; Wei, X.; Kysar, J. W.; Hone, J. Measurement of the elastic properties and intrinsic strength of monolayer graphene. *Science* **2008**, *321*, 385.

19. 19. Dong, X.; Shi, Y.; Zhao Y.; Chen D.; Ye J.; Yao Y.; Gao, F.; Ni, Z.; Yu, T.; Shen, Z.; Huang, Y.; Chen, P.; Li, L. Symmetry breaking of graphene monolayers by molecular decoration. *Phys. Rev. Lett.* **2009**, *102*, 135501.

20. 20. Yavari, F.; Kritzinger, C.; Gaire, C.; Song, L.; Gulapalli, H.; Borca-Tasciuc, T.; Ajayan, P. M.; Koratkar, N. Tunable band gap in graphene by the controlled adsorption of water molecule. *Small* **2010**, *6*, 2535–2538.

21. Rudberg, E.; Salek, P.; Luo, Y. Nonlocal exchange interaction removes half-metallicity in graphene nanoribbons. *Nano Lett.* **2007**, *7*, 2211–2213.

22. Son, Y. W; Cohen, M. L; Louie, S. G. Half-metallic graphene nanoribbons. *Nature* **2006**, *444*, 347.

23. Zhang, Y.; Tang, T.; Girit, C.; Hao, Z.; Martin, M. C.; Zettl, A.; Crommie, M. F., Shen, Y. R.; Wang, F. Direct observation of a widely tunable bandgap in bilayer. *Nature* **2009**, *459*, 820.

24. Zhou, S. Y.; Siegel, D. A.; Fedorov, A. V.; Lanzara, A. Metal to insulator transition in epitaxial graphene induced by molecular doping. *Phys. Rev. Lett.* **2008**, *101*, 086402.

25. Park, S.; Ruoff, R. S. Chemical method for the production of graphene. *Nat. Nanotech* **2009**, *4*, 217–224.

26. Nair, R. R. et al. Fine structure constant defines visual transparency of graphene. *Science* **2008**, *320*, 1308.

27. Bonaccorso, F.; Sun, Z.; Hasan, T; Ferrari, A. C. Graphene photonics and optoelectronics. *Nat. Photonics* **2010**, *4*, 611–620.

28. Roddaro, S.; Pingue, P.; Piazza, V.; Pellegrini, V.; Beltram, F. The optical visibility of graphene: Interference colors of ultrathin graphite on SiO_2. *Nano Lett.* **2007**, *7*, 2707–2710.

29. Chen, Z.; Ren, W.; Gao, L.; Liu, B.; Pei, S.; Chen, H. Three-dimensional flexible and conductive interconnected graphene networks grown by chemical vapour deposition. *Nat. Mater.* **2011**. *10*, 424–428.

30. Li, X.; Zhang, G.; Bai, X.; Sun, X.; Wang, X.; Wang, E.; Dai, H. Highly conducting graphene sheets and Langmuir Blodgett films. *Nat. Nanotech* **2008**, *3*, 538–542.

31. Li, X; Cai, W.; An, J.; Kim S.; Nah, J.; Yang, D.; Piner, R.; Velamakanni, A.; Jung, I.; Tutuc, E.; Banerjee, S. K.; Colombo, L.; Ruoff, R. S. Large-area synthesis of high quality and uniform graphene films on Cu foils. *Science* **2009**, *324*, 1312–1314.

32. Li, X.; Zhu, Y.; Cai, W.; Borysiak, M.; Chen, D.; Piner, R. D.; Colombo, L.; Ruoff, R. S. Transfer of large-area graphene films for high-performance transparent conductive electrodes. *Nano Lett.* **2009**, *9*, 4359–4363.

33. Bae, S.; Kim, H.; Lee, Y.; Xu, X.; Park, J.; Zhen, Y.; Balakrishnan, J.; Lei, T.; Kim, H.; Song, Y.; King, Y.; Kim, K.; Ozyilmaz, B.; Ahn, J.; Hong, B.; Iijima, S. Roll-to-roll production of 30-inch graphene films for transparent electrodes. *Nat. Nanotech* **2010**, *5*, 574–578.

34. Eda, G.; Fanchini, G.; Chhowalla, M. Large-area ultrathin films of reduced graphene oxide as a transparent and flexible electronic material. *Nat. Nanotech* **2008**, *3*, 270–274.

35. Rafiee, J.; Mi, X.; Gullapalli, H.; Thomas, A. V.; Yavari, F.; Shi, Y.; Ajayan, P. M. and Koratkar, N. A. Wetting transparency of graphene. *Nat. Mater.* **2012**, *11*, 217–222.

36. Israelachvili, J. N. Intermolecular and Surface Forces, 2nd Ed: With Applications to Colloidal and Biological Systems; Academic Press, 1992.

37. Seemann, R.; Herminghaus, S.; Jacobs, K. Dewetting patterns and molecular forces: A reconciliation. *Phys. Rev. Lett.* **2001**, *86*, 5534–5537.

38. De Gennes, P. G. Wetting: Statics and dynamics. *Rev. Mod. Phys.* **1985**, *57*, 827–863.

39. Seemann, R.; Herminghaus, S; Jacobs, K. Gaining control of pattern formation of dewetting liquid films. *J. Phys. Condens. Mater.* **2001**, *13*, 4925–4938.

40. Chen, S.; Brown, L.; Levendorf, M.; Cai, W.; Ju, S.; Edgeworth; Li, X.; Magnuson, C.; velamakanni, A.; Piner, R.; Kang, J.; Park, J.; Ruoff, R. S. Oxidation resistance of graphene-coated Cu and Cu/Ni alloy. *ACS Nano* **2011**, *5*, 1321–1327.

41. Cote, L. J.; Cruz-Silva, R.; Huang, J. Flash Reduction and Patterning of Graphite Oxide and Its Polymer Composite. *J. Am. Chem. Soc.* **2009**, *131*, 11027–11032.
42. Zhou, Y.; Bao, Q.; Tang, L. A. L.; Zhong, Y.; Loh, K. P. Hydrothermal dehydration for the "green" reduction of exfoliated graphene oxide to graphene and demonstration of tunable optical limiting properties. *Chem. Mater.* **2009**, *21*, 2950–2956.
43. Ai, K.; Liu, Y.; Lu, L.; Cheng, X.; Huo, L. A novel strategy for making soluble reduced graphene oxide sheets cheaply by adopting an endogenous reducing agent. *J. Mater. Chem.* **2011**, *21*, 3365–3370.
44. Xu, C.; Wang, X.; Zhu, J. Graphene-metal particle nanocomposites. *J. Phys. Chem. C* **2008**, *112*, 19841–19845.
45. Akhavan, O.; Ghaderi, E. Photocatalytic reduction of graphene oxide nanosheets on TiO_2 thin film for photoinactivation of bacteria in solar light irradiation. *J. Phys. Chem. C* **2009**, *113*, 20214–20220.
46. Williams, G.; Seger, B.; Kamat, P. V. TiO2-Graphene Nanocomposites. UV-assisted photocatalytic reduction of graphene oxide. *ACS Nano* **2008**, *2*, 1487–1491.
47. Akhavan, O. Photocatalytic reduction of graphene oxides hybridized by ZnO nanoparticles in ethanol. *Carbon* **2011**, *49*, 11–18.
48. Esfandiar, A.; Akhavan, O.; Irajizad, A. Melatonin as a powerful bio-antioxidant for reduction of graphene oxide. *J. Mater. Chem.* **2011**, *21*, 10907–10914.
49. Akhavan, O.; Ghaderi, E.; Esfandiar, A. Wrapping bacteria by graphene nanosheets for isolation from environment, reactivation by sonication and inactivation by near-infrared irradiation. *J. Phys. Chem. B* **2011**, *115*, 6279–6288.
50. Gao, J.; Liu, F.; Liu, Y.; Ma, N.; Wang, Z.; Zhang, X. Environment-friendly method To produce graphene that employs vitamin C and amino ccid. *Chem. Mater.* **2010**, *22*, 2213–2218.
51. Zhu, C.; Guo, S.; Fang, Y.; Dong, S. Reducing sugar: New functional molecules for the green synthesis of graphene nanosheets. *ACS Nano* **2010**, *4*, 2429–2437.
52. Liu, J.; Fu, S.; Yuan, B.; Li, Y.; Deng, Z. Toward a universal "adhesive nanosheet" for the assembly of multiple nanoparticles based on a protein-induced reduction/decoration of graphene oxide. *J. Am. Chem. Soc.* **2010**, *132*, 7279–7281.
53. Akhavan, O.; Ghaderi, E. E. Escherichia coli bacteria reduce graphene oxide to bactericidal graphene in a self-limiting manner. *Carbon* **2011**, *50*, 1853–1860.
54. Salas, E.; Sun, Z.; Lüttge, A.; Tour, J. M. Reduction of Graphene Oxide via Bacterial Respiration. *ACS Nano* **2010**, *4*, 4852–4856.
55. Schniepp, H. C.; Li, J.-L.; McAllister, M. J.; Sai, H.; Herrera-Alonso, M.; Adamson, D. H.; Prud'homme, R. K.; Car, R.; Saville, D. A.; Aksay, I. A. Functionalized single graphene sheets derived from splitting graphite oxide. *J. Phys. Chem. B* **2006**, *110*, 8535.
56. McAllister, M. J.; Li, J.-L.; Adamson, D. H.; Schniepp, H. C.; Abdala, A. A.; Liu, J.; Herrera-Alonso, M.; Milius, D. L.; Car, R.; Prud'homme, R. K.; Aksay, I. A. Single Sheet Functionalized Graphene by Oxidation and Thermal Expansion of Graphite. *Chem. Mater.* **2007**, *19*, 4396.
57. Rafiee, M.; Rafiee, J.; Yu, Z.; Koratkar, N. Superhydrophobic to superhydrophilic wetting control in graphene films. *Adv. Mater.* **2010**, *22*, 2151–2154
58. Yavari, F.; Rafiee, M. A.; Rafiee, J.; Yu, Z.; Koratkar, N. Dramatic increase in fatigue life in hierarchical graphene composites. *ACS Appl. Mater. Inter.* **2010**, *2*, 2738–2743.
59. Fan, S.; Liu, L.; Liu, M. Monitoring the growth of carbon nanotubes by carbon isotope labelling. *Nanotechnology* **2003**, *14*, 1118.
60. Ferrari, A. C.; Meyer, J. C.; Scardaci, V.; Casiraghi, C.; Lazzeri, M.; Mauri, F.; Piscanec, S.; Jiang, D.; Novoselov, K. S.; Roth, S.; Geim, A. K. Raman spectrum of graphene and graphene layers. *Phys. Rev. Lett.* **2006**, *97*, 187401.
61. Gilje, S.; Han, S.; Wang, M.; Wang, K. L.; Kaner, R. B. A chemical route to graphene for device applications. *Nano Lett.* **2007**, *7*, 3394.
62. Isett, L. C.; Blakely, J. M. Segregation Isosteres for Carbon at the (100) Surface of Nickel. *Surf. Sci.* **1976**, *58*, 397.

63. Li, X.; Cai, W.; An, J.; Kim, S.; Nah, J.; Yang, D.; Piner, R.; Velamakanni, A.; Jung, I.; Tutuc, E.; Banerjee, S. K.; Colombo, L.; Ruoff, R. S. *Science* **2009**, 1171245.

64. Srivastava, A.; Galande, C.; Ci, L.; Song, L.; Rai, C.; Jariwala, D.; Kelly, K.F. and Ajayan, P. M. Novel Liquid precursor-based facile synthesis of large-area continuous, single, and few-layer graphene films. *Chem. Mater.* **2010**, *22*, 3457–3461.

65. Soldano, C.; Mahmood, A.; Dujardin, E. Production, properties and potential of graphene. *Carbon* **2010**, *48*, 2127–2150.

66. Inagaki, M.; Kim, Y. A. and Endo, M. Graphene: preparation and structural perfection. *J. Mater. Chem.* **2010**, DOI: 10.1039/c0jm02991b.

67. Graf, D.; Molitor, F.; Ensslin, K.; Stampfer, C.; Jungen, A.; Hierold, C.; Wirtz, L. Spatially resolved Raman spectroscopy of single- and few-layer graphene. *Nano Lett.* **2007**, *7*, 238–242.

68. Gupta, A.; Chen, G.; Joshi, P.; Tadigadapa, S.; Eklund, P. C. High Frequency Raman scattering from n-graphene layer. *Nano Lett.* **2006**, *6*, 2667–2673.

69. Samuel, J.; Rafiee, J.; Dhiman, P.; Yu Z.-Z: Koratkar, N. Graphene colloidal suspensions as high performance semi-synthetic metal-working fluids. *J. Phys. Chem. B* **2011**, *115*, 3410–3415.

CHAPTER 2

Graphene Polymer Composites: Processing and Characterization of Their Mechanical, Electrical, and Thermal Properties

THIS chapter is aimed at introducing the reader to the processing and the characterization of graphene polymer composites. The resulting nanocomposites show significant enhancement in a variety of mechanical properties, including Young's modulus, tensile strength, buckling stability, fracture toughness, fatigue resistance, creep response, and wear resistance at relatively low nanofiller loading (weight) fractions in comparison to other competing nanofillers such as carbon nanotubes, nanoparticles, and nanoclays. This chapter will discuss in detail the characterization of the mechanical properties of such graphene composites. In addition to mechanical properties, graphene additives are also capable of significantly improving thermal and electrical properties of the polymer and these topics will also be addressed following the discussion of mechanical properties.

The main message to be conveyed in the chapter is that graphene shows outstanding potential for improving the mechanical, thermal, and electrical properties of polymer composites and can provide a performance comparable to other forms of nanofillers at a significantly lower filler loading fraction. This, in addition to their relatively low cost of production (by top down synthesis from graphite), makes graphene strategically well positioned to be the premier nanofiller for a variety of polymer composite applications.

The bulk of the material included in this chapter has been adapted from References [16–18, 43, 51–52, 57–59, 65, 70], published by the author's group and his collaborators.

2.1. PROCESSING AND DISPERSION OF GRAPHENE IN POLYMERS

As discussed in Chapter 1, thermal reduction and exfoliation of graphite oxide can be employed to produce bulk quantities of graphene platelets (GPLs). Briefly, one-step reduction and exfoliation of graphite oxide into GPLs [1–3] can be achieved by exposing the graphite oxide powder to a thermal shock (heating rate of ~2000°C/min). Because of the rapid rate of heating, the oxygen groups expelled primarily as CO_2 gas are unable to diffuse out of the layered graphite structure. This results in a large pressure build-up that overcomes the interlayer Van der Waals forces and exfoliates the graphite structure into GPLs. Here, the term platelet instead of sheet is used to specify that a GPL is not a single-/bilayer graphene. Hence, the GPL's thickness is larger than the monolayer graphene (i.e., 0.34 nm) and is made of few-layer graphene (1–2 nm thick).

There are a variety of techniques that can be used to achieve uniform dispersion of graphene in polymer matrices. In this chapter, we will mainly focus our attention on epoxies as they are the extensively used in structural applications ranging from floor coatings to aircraft fuselages. Due to low shrinkage and a multitude of available curing options, epoxy is one of the most commonly used adhesives [4–5]. Its major application is in the field of structural metal bonding in the aerospace industry, in addition to small-part assemblies. They also make excellent coating materials due to their high corrosion resistance, strong adhesion, and good physical properties. Bisphenol A products (compounds with two phenol functional groups) are the most commonly used epoxy coatings. Epoxies are also a commonly used encapsulant for circuit components because the application requires high electrical insulation and isolation from adverse environmental conditions of temperature, moisture, and chemicals. Epoxy polymers are also extensively used as a stabilizer for plastics.

For thermosetting polymers such as epoxies, solution mixing (via ultrasonication), followed by high-speed shear mixing, is the preferred option for graphene dispersion. Solution mixing is far more effective at achieving uniform graphene dispersion compared to conventional melt mixing [6–7]. In this method, the required quantity of GPL is first dispersed in an acetone solution (~100 mL acetone for 0.1 g GPL) via high amplitude ultrasonic (~75 Hz) sonication technique (Sonics Vibracell VC 750, Sonics and Materials Inc., USA) for ~1.5 h in an ice bath (temperature ~10°C). The epoxy monomer can then be added to the solution and further sonicated following the same protocol for ~1.5 h. Next, the acetone is gradually removed through heating the mixture on a stir magnetic plate for ~3 h at 70°C. Subsequently, the solution is placed in a vacuum environment (–30 in Hg) for ~12 h at 70°C. After cooling the GPL/epoxy mixture to room temperature (~23°C), a low-viscosity curing agent is added, and a high-speed shear mixer (e.g., ARE-250, Thinky, Japan) is used to mix the GPL/epoxy slurry for ~4 min at ~2000 rpm. Then, in order to remove air bubbles, the mixture is placed in a vacuum environment (–30 in Hg) for ~30 min. Finally, silicone molds are used to prepare the nanocomposite samples, which are typically cured at room temperature (~23°C) and under pressure (90 psi) for ~24 h, and post cured for several hours at elevated temperatures (~90°C). Figure 2.1 is a typical schematic representation of the cure process. In this chapter, I have focused on one specific epoxy monomer (System 2000 Epoxy Resin, Fibreglast Inc., USA) and curing agent (2120 Epoxy Hardener, Fibreglast Inc., USA) combination. Such bisphenol A-based epoxies are among the most widely used polymers for structural applications. In addition to thermosetting epoxies, I will also present

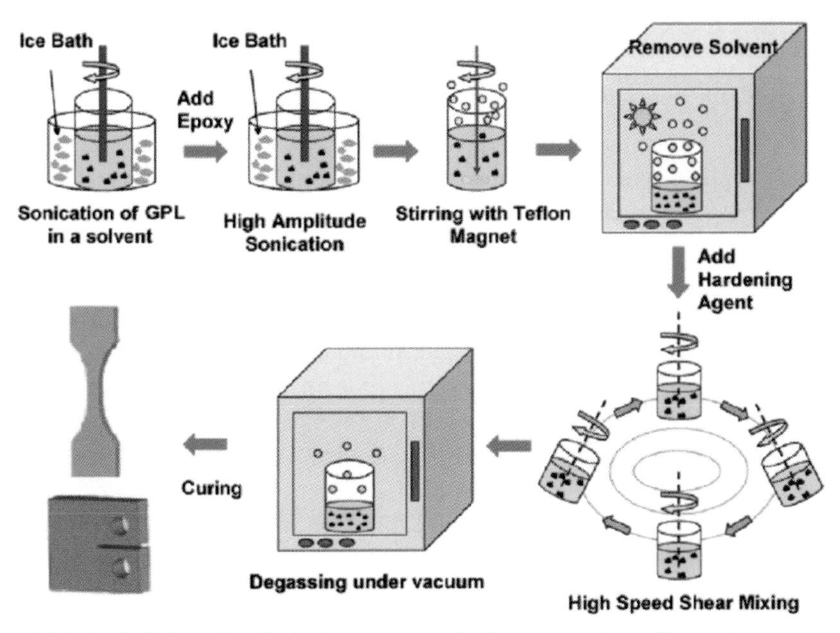

FIGURE 2.1. Schematic of typical process used to disperse graphene fillers in thermosetting epoxy polymers. (Adapted from [18] with permission).

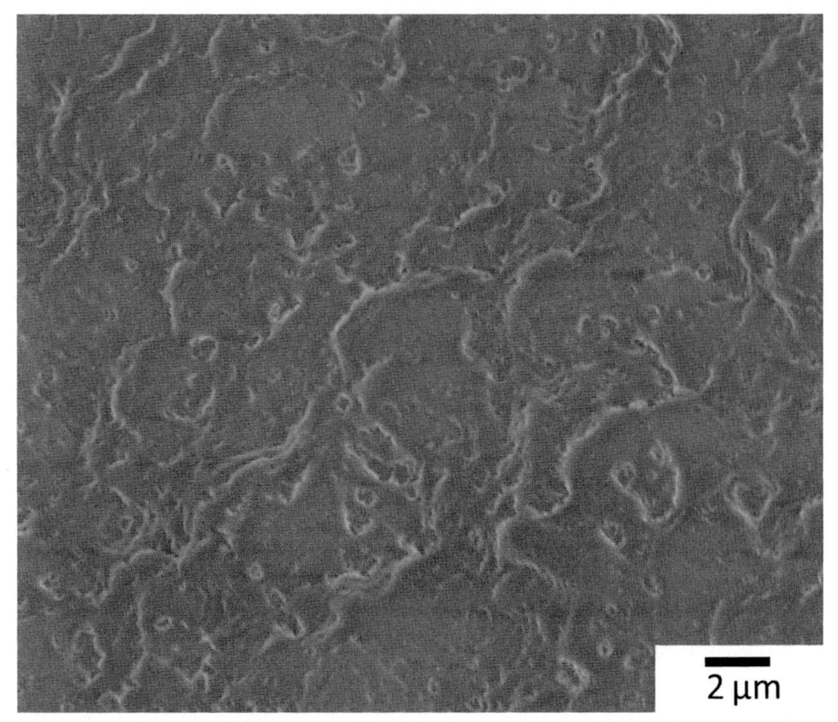

FIGURE 2.2. *Scanning electron microscopy (SEM) image of the freeze-fractured surface of a graphene/epoxy composite with ~0.5% weight fraction of graphene platelets (GPLs). The platelets tend to agglomerate into clusters. The wavy polymer smeared graphene edges are discernable from the image.*

some typical results for commonly used thermoplastic polymers with graphene reinforcement.

Scanning electron microscopy (SEM) and optical microscopy are the commonly used methods to study graphene dispersion in polymer composites [8–9]. SEM is performed after freeze-fracturing the bulk composite sample using liquid nitrogen. In the case of optical microscopy, it is important to have a smooth surface for imaging and, therefore, thin microtomed slices of the composite are typically utilized. Figure 2.2 shows an SEM image of a freeze-fractured nanocomposite with ~0.5% weight of GPL. The polymer used here is a thermosetting epoxy (Epoxy 2000 from Fibreglast, USA). The image clearly indicates the wavy edges of epoxy-coated GPL on the fracture surface of the sample. There appears to be significant clustering (agglomeration) of the GPL; however, the epoxy does appear to wet the sheet edges. Figure 2.3 shows an SEM image of the freeze-fractured surface of an epoxy polymer with ~0.1% weight fraction of GPL. The image shows a single isolated graphene platelet that is embedded in the epoxy matrix. There is no indication of debonding or pull-out of the graphene sheet, which suggests a strong interface. Note the formation of wrinkles on the surface of the platelet. Such wrinkles can play an important role in improving mechanical interlocking [10] with the surrounding polymer matrix.

Optical micrographs [8] can also be used to study graphene dispersion. Such micrographs are typically obtained in the transmission mode from thin microtomed slices of the composite material. Figure 2.4 shows a typical optical micrograph of an epoxy composite with ~0.1% weight fraction of GPL additives. The micrometer scale size (lateral dimensions) of the GPL makes it possible to observe the graphene dispersion at low magnifications using an optical microscope. To study the onset of agglomeration in the polymer matrix, the average GPL particle size can be studied as a functional of GPL loading fraction. For this, optical micrographs of translucent GPL-epoxy nanocomposite samples prepared by microtoming can be obtained

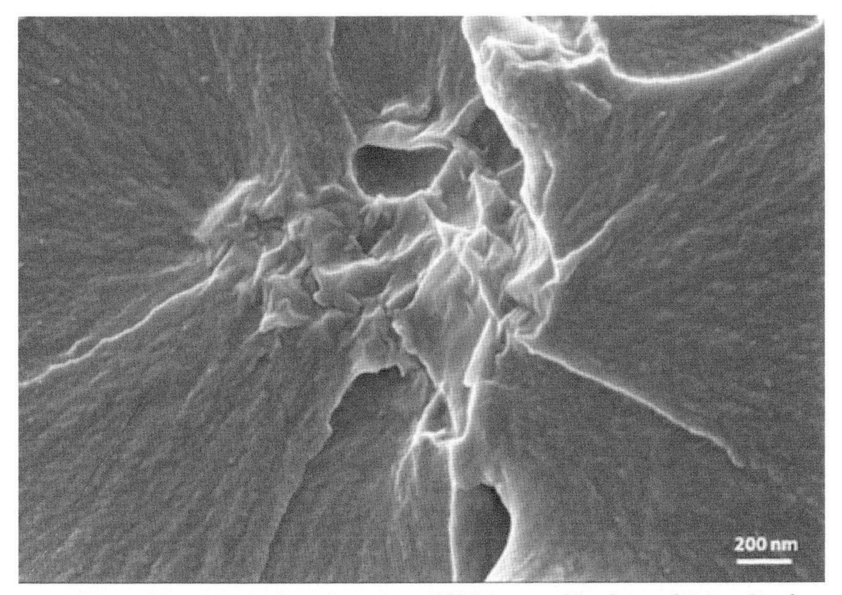

FIGURE 2.3. *Scanning electron microscopy (SEM) image of the freeze-fractured surface of a graphene/epoxy composite with ~0.1% weight fraction of graphene platelets (GPLs). The image shows a typical platelet (~2 μm × 2 μm micron in size) embedded in the epoxy matrix. The epoxy appears to strongly wet the graphene surface, which suggests a strong interface. (Adapted from [51] with permission).*

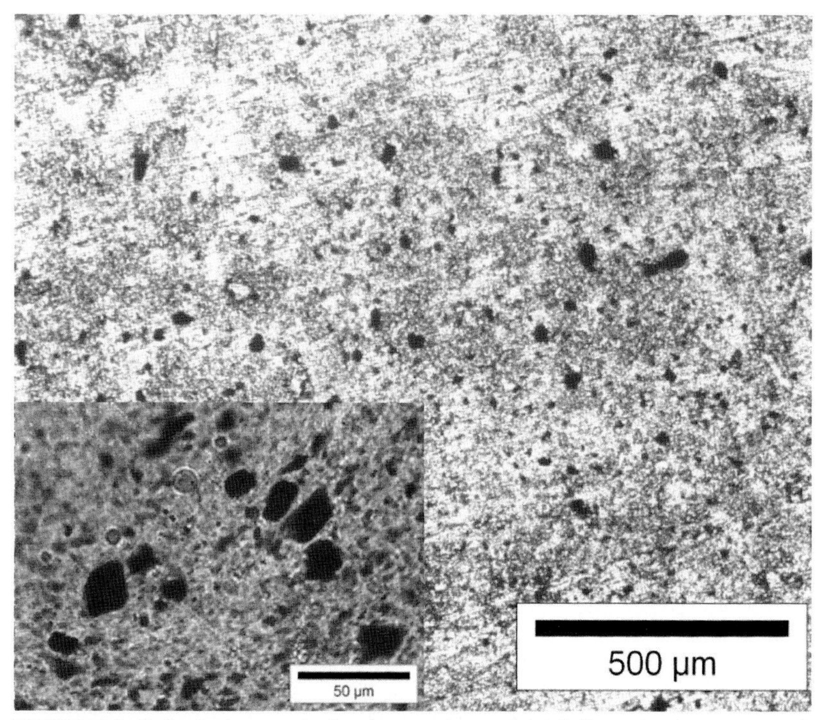

FIGURE 2.4. Optical micrograph showing graphene platelet dispersion in an epoxy. In the inset, the average GPL diameter can be seen to be ~5 μm, scale bar (50 μm).

under transmittance mode. These micrographs can be converted into binary images for image analysis as shown in Figure 2.5(a). Figure 2.5(b) shows the typical plot of average graphene platelet size (diameter) as a function of its weight % for an epoxy polymer. At each GPL weight %, at least 4–5 samples should be imaged to obtain a reliable estimate of the average platelet diameter. It is clear from the plot that there is a sharp increase in the graphene cluster size above a loading fraction of ~0.125%, indicating a significant degradation in dispersion above ~0.125% weight fraction of GPLs.

2.2. TENSILE PROPERTIES: YOUNG'S MODULUS AND ULTIMATE TENSILE STRENGTH

Tensile properties of a material can be determined from stress vs. strain curves obtained by the application of a static load uniformly over the material cross-section. Tensile tests are commonly done to obtain the relevant mechanical properties of a material, such as Young's modu-

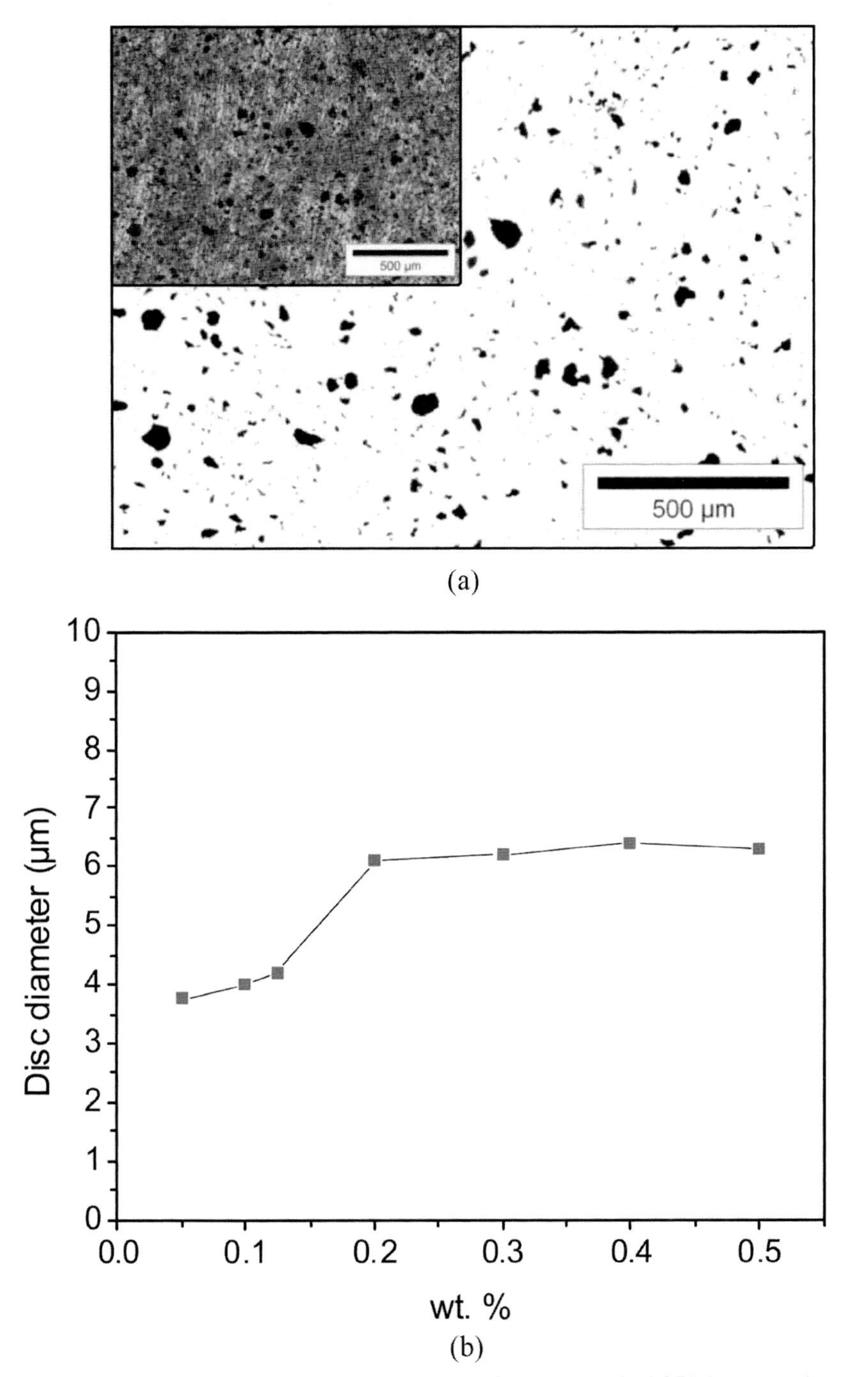

FIGURE 2.5. *(a) Low magnification binary image of 0.1 weight % of GPL in epoxy; the inset shows the actual optical micrograph. (b) Graphene platelet diameter as a function of its increasing weight % in epoxy.*

lus, strain-to-failure, ultimate tensile strength, and toughness (i.e., total energy absorbed prior to failure).

The engineering stress and engineering strain in the material are used to avoid the use of a geometry factor. The engineering stress σ is given by:

$$\sigma = \frac{F}{A} \tag{1}$$

where F is the instantaneous load applied perpendicular to the sample and A is the original cross-sectional area of the sample. The ultimate tensile strength (UTS) is the maximum stress a material can withstand under tensile load. It is given by the maximum stress in the engineering stress-strain curve. The engineering strain σ is given by:

$$\varepsilon = \frac{L - L_o}{L_o} \tag{2}$$

where L is the final sample length and L_o is the original sample length.

Young's modulus measures the ability of a material to resist elastic deformation under applied stress, given by the slope of the stress-strain curve below the yield point. It is given by Hooke's law between σ and ε:

$$E = \frac{\sigma}{\varepsilon} \tag{3}$$

The modulus is a direct measure of microscopic interatomic resistance to stretching. Different materials have different modulus, depending on their interatomic bonding. Polymers have a range of modulus, varying from a few MPa (e.g., Polydimethylsiloxane) up to 3 GPa (e.g., epoxy resins). Compared to polymers, the modulus values for ceramics reside at the other end of the spectrum, varying between a few tens to few hundreds of GPa. The two materials with highest reported Young's modulus are carbon nanotubes and graphene, with values approaching 1 TPa [10].

Various theoretical models in the literature and their modifications complemented with experimental results help predict and estimate the composite modulus as a function of matrix, filler, and interface properties. Some of the models are discussed below.

Iso-stress and Iso-strain model: The composite modulus generally lies between the upper and lower bound of the modulus given by the Iso-strain [Equation (4)] and the Iso-stress [Equation (5)] models [4].

$$E_c = E_f V_f + E_m V_m \tag{4}$$

$$E_c = \frac{E_f E_m}{E_f V_m + E_m V_f} \tag{5}$$

where E_m, E_f and E_c are the elastic modulus of the matrix, filler, and composite, respectively, and V_m and V_f are the volume fractions of matrix and filler. Equations (4) and (5), called the rule of mixture and the inverse rule of mixture, give poor modulus prediction as they assume all the fillers have the same orientation and do not consider imperfect filler-matrix interface properties.

Mori-Tanaka model: The Mori-Tanaka model is used to predict the stiffness tensor of composites under the assumption that the fibers and matrix undergo the same average strain. Derived from the principles of Eshelby's inclusion model for ellipsoidal particles in an infinite matrix, this model gives the best predictions for high aspect ratio fillers, but predicts negligible reinforcement for aspect ratio equal to unity, i.e., particulate fillers. The model [4] effectively takes into account the macroscopic average strain and the fluctuation strain due to the neighboring fibers, giving the stiffness of the composite as:

$$\begin{aligned}\mathbf{C}^{comp} = \mathbf{C}^m + f_f(\mathbf{C}^f - \mathbf{C}^m)[\mathbf{I} + \mathbf{S}^E(\mathbf{C}^m)^{-1}(\mathbf{C}^f - \mathbf{C}^m)]^{-1} * \\ (f_m\mathbf{I} + f_f[\mathbf{I} + \mathbf{S}^E(\mathbf{C}^m)^{-1}(\mathbf{C}^f - \mathbf{C}^m)]^{-1})^{-1}\end{aligned} \tag{6}$$

where, $\mathbf{C}^{comp}$, $\mathbf{C}^m$ and $\mathbf{C}^f$ are the stiffness matrices of the composite, matrix, and fiber , respectively; $\mathbf{S}^{comp}$, $\mathbf{S}^m$ and $\mathbf{S}^f$ are the compliance matrices of the composite, matrix, and fiber, respectively; and f_f and f_m are the volume fractions of fiber and matrix in the composite and $\mathbf{I}$ is a unit matrix.

Halpin-Tsai model: Another very useful and practical model for uni-directional and discontinuous-filler composites is the Halpin-Tsai model [11]. The Halpin-Tsai equations are useful for obtaining various composite properties, including modulus, using a parameter ζ, which is a measure of the reinforcement geometry and loading conditions. The modulus of the composite is given by:

$$E_c = E_m\left(\frac{1 + \varsigma\eta V_f}{1 - \eta V_f}\right) \tag{7}$$

$$\eta = \frac{\frac{E_f}{E_m} - 1}{\frac{E_f}{E_m} + \varsigma} \tag{8}$$

where E_m, E_f and E_c are the elastic modulus of the matrix, filler, and composite, respectively, and V_f is the filler volume fraction. The Halpin-Tsai equation [Equation (7)] has particularly proven very useful for predicting the properties of short-fiber and particulate-reinforced composites. For $\zeta = 0$ the equation takes the form of the inverse rule of mixture and for $\zeta = \infty$ it takes the form of the rule of mixture. A shape factor of $2w/t$ (w is filler length, t is filler thickness) is taken for ζ [12], which reduces for a spherical particle to $\zeta = 2$.

As Halpin-Tsai considered only the ideal case of a perfect particle-matrix adhesion, the model was improved by Lewis-Nielsen and McGee and McCullough [13–14], whose models take into account weak filler-matrix interfaces. The modified composite modulus was given as:

$$E_c = E_m \left\{ \frac{1 + (k_E - 1)\beta V_f}{1 - \beta\mu V_f} \right\} \tag{9}$$

where k_E is the Einstein's coefficient, which determines the degree of filler-matrix adhesion. β takes into account the relative modulus of the filler and the matrix. It is given by:

$$\beta = \frac{\frac{E_f}{E_m} - 1}{\frac{E_f}{E_m} + (k_E - 1)} \tag{10}$$

and μ is given as:

$$\mu = 1 + \frac{(1 - V_f)}{V_{max}} [V_{max} V_f + (1 - V_{max}) V_m] \tag{11}$$

where V_{max} is calculated from the Nielsen and Landel model [15].

Static tensile tests are typically conducted on dog-bone shaped [Figure 2.6(a)] test coupons following ASTM D638 standard. Figure

2.6(b) shows typical tensile stress vs. tensile strain plots for GPL/epoxy nanocomposites with ~0.1% of GPL additives. For comparison, data is also shown for the same weight % of single-walled carbon nanotubes (SWNTs) and multi-walled carbon nanotubes (MWNTs) in epoxy, as well as the neat (pristine) epoxy matrix. The results of the static tensile tests [16–17] are summarized in Figure 2.7. To check for repeatability of the data, at least four specimens of each nanocomposite at the same filler weight fraction of ~0.1 wt. % were manufactured and tested. The weight fraction of 0.1% was chosen to ensure relatively homogeneous dispersion of fillers. Figure 2.7(a) shows the ultimate tensile strength measurements for the pure epoxy and the nanocomposite samples. Clearly, the GPL additives far outperform the SWNT and MWNT fillers. The tensile strength of the GPL/epoxy nanocomposite (~78 MPa) is

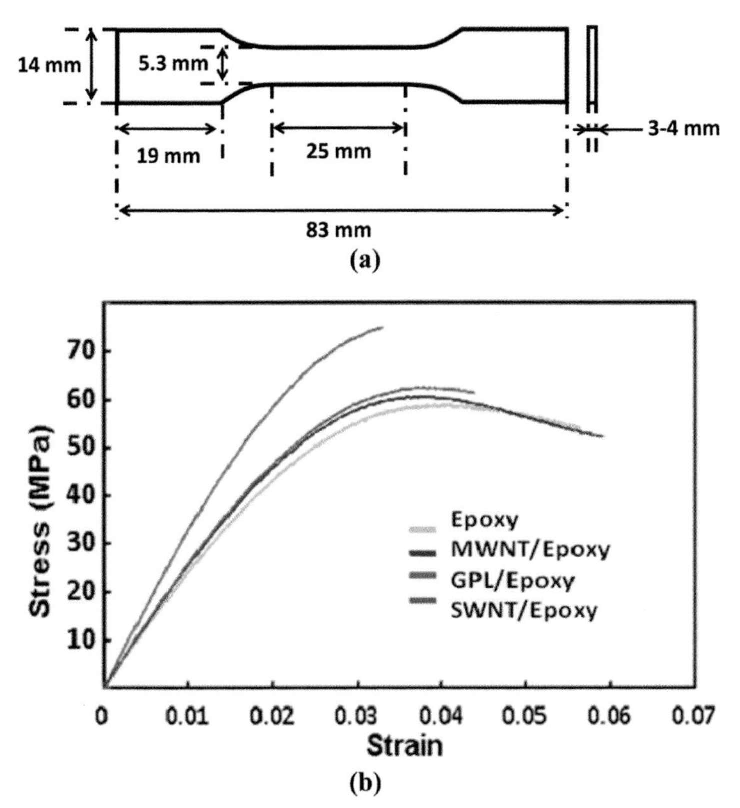

FIGURE 2.6. *(a) Sample geometry for tensile tests. (b) Typical tensile stress vs. tensile strain curves for testing of baseline epoxy and nanocomposites with ~0.1% weight fraction of graphene (GPL), multi-walled carbon nanotube (MWNT), and single-walled carbon nanotube (SWNT) fillers. (Adapted from [16–17] with permission).*

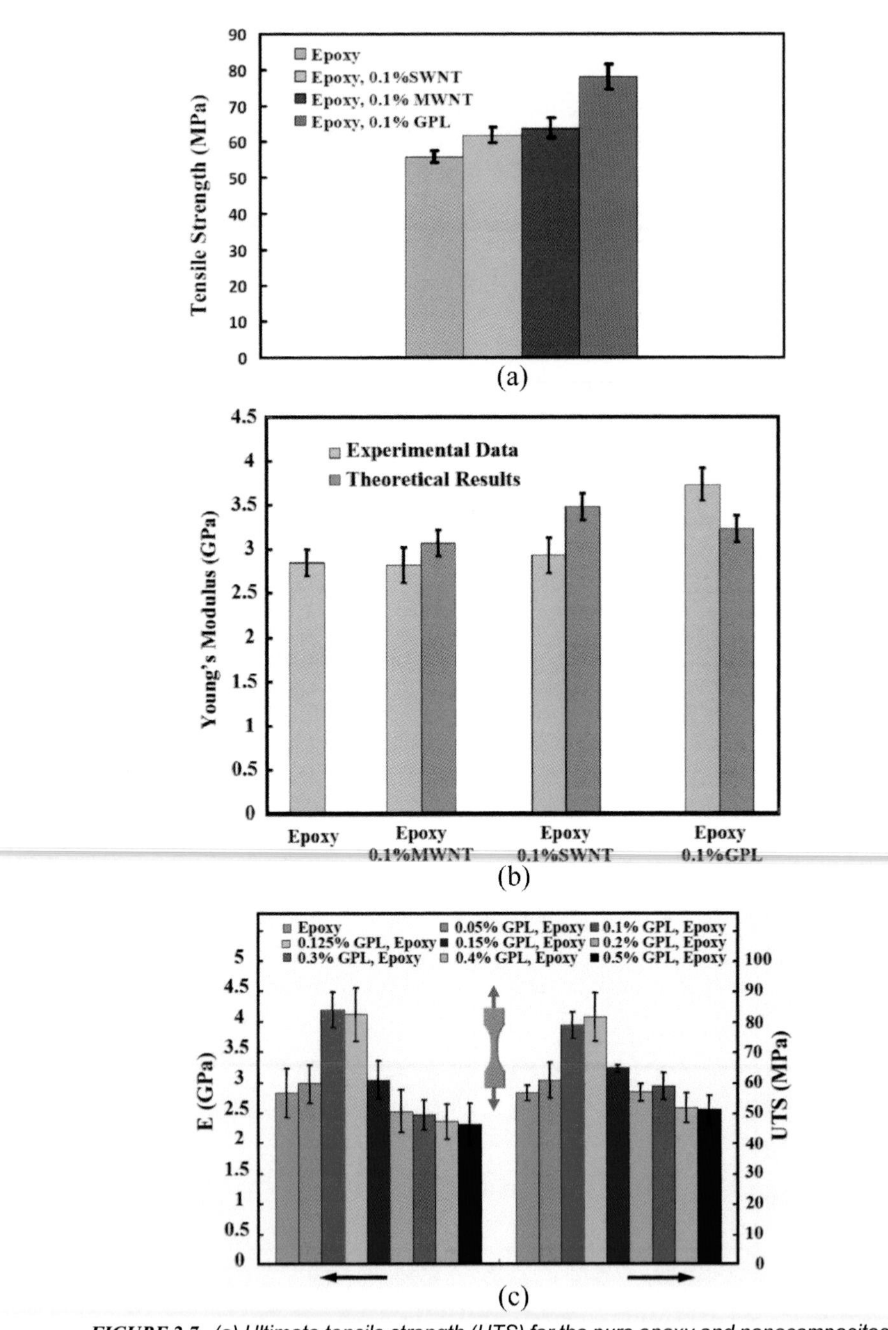

FIGURE 2.7. *(a) Ultimate tensile strength (UTS) for the pure epoxy and nanocomposites. (b) Young's modulus (E) of nanocomposite samples with 0.1 wt. % weight of fillers is compared to the baseline epoxy. Theoretical results computed using the Halpin-Tsai equation for fiber-reinforced composite materials are also shown. (c) Effect of GPL loading fraction on the Young's modulus (E) and ultimate tensile strength (UTS) of the composite structure. (Adapted from [16–17] with permission).*

about 40% larger than the pristine epoxy (~55 MPa). The fact that this is achieved at a nanofiller weight fraction of only ~0.1% is impressive. By contrast, SWNT/epoxy and MWNT/epoxy composites show ~11 and ~14% increases, respectively, in the tensile strength compared to the baseline epoxy matrix. Figure 2.7(b) compares the Young's modulus of the pristine epoxy and nanocomposite samples. Incorporation of ~0.1 weight fraction of GPL increases the Young's modulus of the baseline epoxy by ~31% from ~2.85 to ~3.74 GPa. The modulus enhancements for SWNT and MWNT composites at the same weight fraction of ~0.1% are significantly lower (< 3%). Figure 2.7(c) quantifies the effect of GPL loading fraction on the Young's modulus (E) and ultimate tensile strength (UTS) of the composite. The optimal GPL loading fraction is ~0.125%, after which the E and UTS begin to deteriorate back to the pure epoxy value. This is likely due to degradation in dispersion of the GPL at loading fractions above ~0.125%, which is also confirmed by the optical microscopy study of average graphene particle size shown previously in Figure 2.5(b).

There are three main reasons for the superiority of graphene over carbon nanotubes in terms of improving the tensile modulus and strength of composites:

(1) *Surface Area:* For carbon nanotubes, only the outer surface of the tube is in contact with the matrix. This is because polymer chains are typically too large to penetrate into the inner pores of nanotubes. By contrast, for a flat sheet such as graphene, both surfaces of the sheet can interface with the polymer. This leads to greater interfacial contact area with the polymer.

(2) *Interface:* Graphene shows a rough and wrinkled (wavy) surface topology, which can enable strong mechanical interlocking with polymer chains. This is in contrast to the atomistically smooth surfaces of carbon nanotubes that hinder interlocking with the encapsulating polymer matrix.

(3) *Orientation:* One-dimensional (1D) high aspect ratio fibers, such as carbon nanotubes, can carry load only along the longitudinal axis of the fiber. This places severe constraints on the alignment of the fiber with respect to the loading direction. In a randomly oriented nanotube composites, only a small fraction of the nanotubes will be well aligned with the loading direction and, therefore, able to participate in reinforcing the polymer. In the case of a two-dimensional (2D) filler such as graphene, the fiber is able

to support load in both the longitudinal and the lateral directions. Therefore, in a composite with randomly oriented nanofillers, the effectiveness of graphene as a structural reinforcement additive will be superior to that of carbon nanotubes.

It is important to explore mathematical models to predict the mechanical properties of graphene nanocomposites. The Halpin-Tsai theory for fiber-reinforced composites is a well-established model [18] that can be utilized to predict the elastic properties of nanocomposite materials. Carbon nanotube (CNT) composites resemble a randomly oriented discontinuous fiber network. Halpin-Tsai equations applied to such a system yield the following result for the Young's modulus (E):

$$E_c = \left(\frac{3}{8} E_M \frac{1 + 2(l_{NT}/d_{NT}) \left(\dfrac{(E_{eq}/E_M) - 1}{(E_{eq}/E_M) + 2(l_{NT}/d_{NT})} \right) V_{NT}}{1 - \left(\dfrac{(E_{eq}/E_M) - 1}{(E_{eq}/E_M) + 2(l_{NT}/d_{NT})} \right) V_{NT}} + \frac{5}{8} E_M \frac{1 + 2 \left(\dfrac{(E_{eq}/E_M) - 1}{(E_{eq}/E_M) + 2} \right) V_{NT}}{1 - \left(\dfrac{(E_{eq}/E_M) - 1}{(E_{eq}/E_M) + 2} \right) V_{NT}} \right) \quad (12)$$

where E_C = elastic modulus of the composite, l_{NT} = CNT length (l_{SWNT} = 10 μm, l_{MWNT} = 20 μm), d_{NT} = CNT average diameter (d_{SWNT} = 2 nm, d_{SWNT} = 20 nm), E_{NT} = elastic modulus of the CNTs (E_{SWNT} = 1 TPa, E_{MWNT} = 450 GPa), E_M = elastic modulus of the pristine epoxy, E_{eq} = $[(2t)/(r^{NT})]E_{NT}$ is the equivalent modulus of the nanotube considering the hollow tube as a solid cylinder, t = CNT wall thickness (t_{SWNT} = 0.34 nm, t_{MWNT} = 1.5 nm), r^{NT} = CNT radius (r^{SWNT} = 1 nm, r^{MWNT} = 10 nm), and V_{NT} = volume fraction of the CNTs (V_{SWNT} = 0.171 vol.%, V_{MWNT} = 0.138 vol.%). The CNT (both SWNTs and MWNTs) density can be estimated based on the known graphite density (2.25 gr/cm^3) and the supplied CNT diameter provided by the manufacturers ($d_{outer,SWNT}$ = 2 nm, $d_{inner,SWNT}$ = 1.66 nm, $d_{outer,MWNT}$ = 20 nm, $d_{inner,MWNT}$ = 17 nm). The pristine epoxy density is typically ~1.2 g/cm^3.

To model the modulus (E) of the graphene-reinforced nanocomposites, the GPL can be considered to be acting as an effective solid filler (rectangular shaped) with width (W), length (L), and thickness (t). To calculate the elastic properties, the Halpin-Tsai equations [18] can be modified for the GPL-based nanocomposites as follows:

$$E_c = \frac{3}{8} E_M \frac{1 + \xi \eta_L Ve_{ff,fib}}{1 - \eta_L Ve_{ff,fib}} + \frac{5}{8} E_M \frac{1 + \eta_W Ve_{ff,fib}}{1 - \eta_W Ve_{ff,fib}} \tag{13}$$

$$\eta_L = \frac{(E_{eff,fib}/E_M) - 1}{(E_{eff,fib}/E_M) + \xi} \tag{14}$$

$$\eta_W = \frac{(E_{eff,fib}/E_M) - 1}{(E_{eff,fib}/E_M) + 2} \tag{15}$$

where E_C is the Young's modulus of the nanocomposite, $V_{eff,fib}$ is effective filler volume fraction; $E_{eff,fib}$ and E_M are the effective filler and matrix moduli. $E_{eff,fib}$ can be assumed to be similar to the GPL modulus (~1.01 TPa). The parameter ξ is the geometry factor of the effective filler. According to the Halpin-Thomas theory for rectangular-shaped fillers, the geometry factor can be assumed as:

$$\xi = 2\left(\frac{(W + L)/2}{t}\right) \tag{16}$$

where L, W, and t are the average length, width, and thickness of the GPL. In this way, the Young's modulus of the nanocomposite can be defined in terms of the epoxy and the GPL properties. Assuming that $V_{eff,fib} = V_{\text{GPL}}$ and by substituting Equations (14), (15), and (16) into Equation (13), we finally obtain:

$$E_c = \left(\frac{3}{8} E_M \frac{1 + ((W+L)/t)\left(\frac{(E_{GPL}/E_M) - 1}{(E_{GPL}/E_M) + (W+L)/t)}\right) V_{GPL}}{1 - \left(\frac{(E_{GPL}/E_M) - 1}{(E_{GPL}/E_M) + (W+L)/t)}\right) V_{GPL}} + \frac{5}{8} E_M \frac{1 + 2\left(\frac{(E_{GPL}/E_M) - 1}{(E_{GPL}/E_M) + 2}\right) V_{GPL}}{1 - \left(\frac{(E_{GPL}/E_M) - 1}{(E_{GPL}/E_M) + 2}\right) V_{GPL}} \right) \tag{17}$$

where, E_C is the Young's modulus of the GPL nanocomposite. The density of the GPL and epoxy should be known in order to convert weight fraction to volume fraction, required to predict the elastic properties. For fibrous composites, the fiber volume fraction can be calculated using the density of the constituents:

$$V_{GPL} = \frac{\rho_C}{\rho_{GPL}} W_{GPL} \qquad \rho_C = \rho_{GPL} V_{GPL} + \rho_M V_M \tag{18}$$

where V_{GPL} and W_{GPL} are the vol. % and wt. % of the GPL, V_M is the vol. % of the pure matrix, ρ_C is the nanocomposite density, ρ_{GPL} is the GPL density, and ρ_M is the density of the matrix. By rearranging Equation (18), the volume fraction can be expressed as follows:

$$V_{GPL} = \frac{W_{GPL}}{W_{GPL} + (\rho_{GPL} + \rho_M)(1 - W_{GPL})} \tag{19}$$

Chapter 1 showed high resolution transmission electron micrographs (HRTEM) of the edge of a GPL showing the layered platelet structure. The interlayer spacing ($t' \approx 0.72$ nm) of the GPLs can be directly measured from such micrographs by image processing techniques. This enables the GPL density to be estimated by appropriately scaling the density of fully dense graphite ($\rho_{graphite} \approx 2.25$ g/cm^3), with $t' \approx 0.34$ nm as the interlayer spacing in graphite. Utilizing this procedure, the GPL density can be approximated as ~1.06 gr/cm^3. The projected volume fraction for 0.1% weight of GPL then calculates to $V_{GPL} = 0.112$ vol.%. The theoretical prediction from Equation (6) (with L $\approx$ 2.5 μm, $W \approx 1.5$ μm, $t \approx 1.5$ nm) is ~3.23 GPa [Figure 2.7(b)], which under-predicts the experimental results by ~13%. This could be due to the wrinkled (wavy) structure of the GPL, which is different from the rectangular cross- sectional shape of the GPL assumed by the model. The calculations of the Halpin-Tsai model for MWNT and SWNT epoxy nanocomposites are also shown in Figure 2.7(b). For the carbon nanotube composites, the theory over-predicts the test data by ~12%.

2.3. COMPRESSIVE PROPERTIES: BUCKLING STABILITY

The enhanced elastic modulus of graphene composites over the baseline epoxy also suggests an improvement in buckling stability [17] under compressive loading. Buckling is a structural instability failure mode and a major concern for structural design. Buckling is related to both the geometry and the material properties of the structure. For a slender column under compression, buckling usually occurs well before the allowable normal stress of the material is reached. For a column under an axial compressive load, the smallest critical load that defines the onset of structural instability is given by Euler's equation:

$$P_{Buckling} = \frac{\pi^2 EI}{L_e^2} \tag{20}$$

where $P_{Buckling}$ is the critical buckling load, E is the elastic modulus of the column, L_e is the effective length of the column, and I is the moment of inertia of the cross section. The effective length L_e depends on the column boundary conditions. For fixed boundary conditions, the effective length is half of the gage length of the column. The specimens tested in this section have a slenderness ratio $L_e/\rho \approx 45$ (where ρ is the column radius), greater than the critical slenderness ratio $SR_c \approx 40$, which means that the column can be considered to be long and Euler's equation can be utilized. The slenderness ratio and the critical slenderness ratio are computed using:

$$\frac{L_e}{\rho} = \frac{L_{gage}/2}{\sqrt{I/A}} \tag{21}$$

$$SR_c = \sqrt{\frac{E\pi^2}{\sigma_{pl}}} \tag{22}$$

where L_{gage} is the gage length, I is the least moment of inertia of the cross section, A is the area of the cross section, and σ_{pl} is the proportional limit of the material. From the classical Euler equation, it is clear that addition of graphene reinforcement into the matrix material will increase the elastic modulus of the sample, causing a corresponding increase in the critical buckling load. Therefore, the buckling stability enhancement is expected to be proportional to the stiffening (i.e., the elastic modulus enhancement) of the composite structure.

The baseline epoxy and nanocomposite samples tested in this section are ~90–100 mm in clamped length, ~24.5 mm in width, and ~3.5–3.9 mm in thickness [17]. In addition to the GPL/epoxy composites, test data for single-walled carbon nanotube (SWNT) and multi-walled carbon nanotube (MWNT) epoxy composites are also presented to compare the performance of carbon nanotube and graphene additives. The same protocols were used for SWNT and MWNT dispersion in the epoxy matrix. Note that, in general, carbon nanotubes are relatively easier to disperse than GPL, since the 2D sheet geometry and micron scale lateral dimensions of GPL increase the probability of nanofiller entanglement and clustering. Therefore, to ensure relatively uniform dispersion for all the three nanofiller systems (i.e., MWNT, SWNT, and GPL), a low nanofiller weight fraction of ~0.001 (i.e., 0.1%) was used in the experiments.

The tensile elastic modulus of the samples was measured prior to the buckling tests. An extensometer was attached to the specimen to measure the strain during the tests. The Young's moduli of the nanocomposites [Figure 2.8(b)] were calculated from the slope of the stress-strain curves [Figure 2.8(a)]. There was only ~0.5% increase in the Young's modulus of the 0.1% weight MWNT nanocomposite compared to the neat epoxy, while the Young's modulus of the 0.1% weight SWNT and 0.1% weight GPL nanocomposites showed approximately ~4% and ~32% increases, respectively. Note that when computing the theoretical buckling load, the measured tensile modulus is typically used. The compressive modulus of each test specimen on initial compressive loading (up to –0.1% strain) was also monitored and the measured compressive modulus was very close to the tensile modulus values (maximum difference of only ±1.5%). The compressive modulus is typically not measured at the higher compressive strains since the bowing of the specimen results in inaccurate strain measurement from the extensometer.

Having measured the elastic modulus of nanocomposites, the samples were then buckled by the application of a monotonically increasing compressive displacement to the specimen (at the rate of ~0.1 mm/min). The resulting typical load–displacement response [Figure 2.9(a)] was used to determine the buckling load. At the point of buckling, the system is unstable, with the displacement continuing to increase without any further increase of the load (i.e., the load response levels off). The measured buckling load from Figure 2.9(a) has been scaled appropriately using:

$$P_{Buckling,Scaled} = P_{Buckling} \frac{L^2}{L_{ref}^2} \frac{I_{ref}}{I} \tag{23}$$

where $P_{Buckling,Scaled}$ is the scaled buckling load used for comparison, $P_{Buckling}$ is the buckling load obtained from the experiment, L is the clamped length of the nanocomposite, I is the moment of inertia of the nanocomposite, L_{ref} is the clamped length of the reference (pure epoxy), and I_{ref} is the moment of inertia of the reference (pure epoxy) samples. The scaling of the buckling load [Equation (23)] with the epoxy beam dimensions taken as a reference allows us to directly compare the buckling loads of the specimens. This is necessary to account for slight variations in the clamped length, thickness, and width of the samples. For

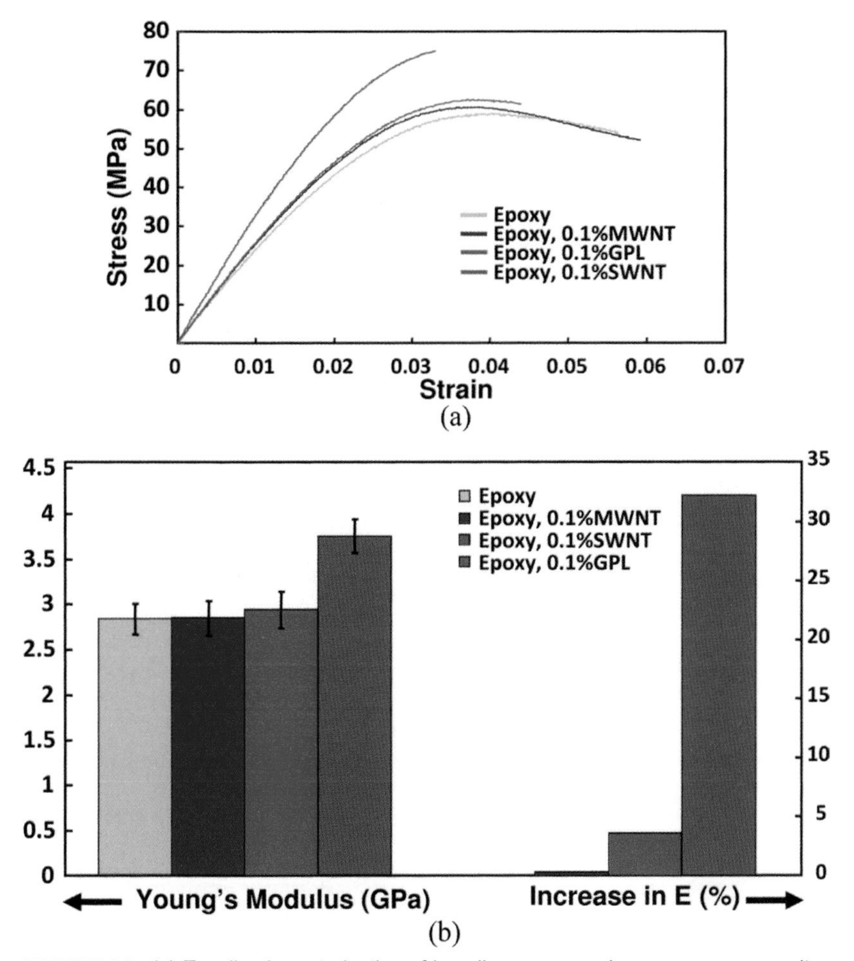

FIGURE 2.8. (a) Tensile characterization of baseline epoxy and epoxy nanocomposites with 0.1% of GPL, MWNT, and SWNT additives respectively. (b) Absolute and percentage increase in Young's modulus for the baseline epoxy and nanocomposite samples respectively. (Adapted from [17] with permission).

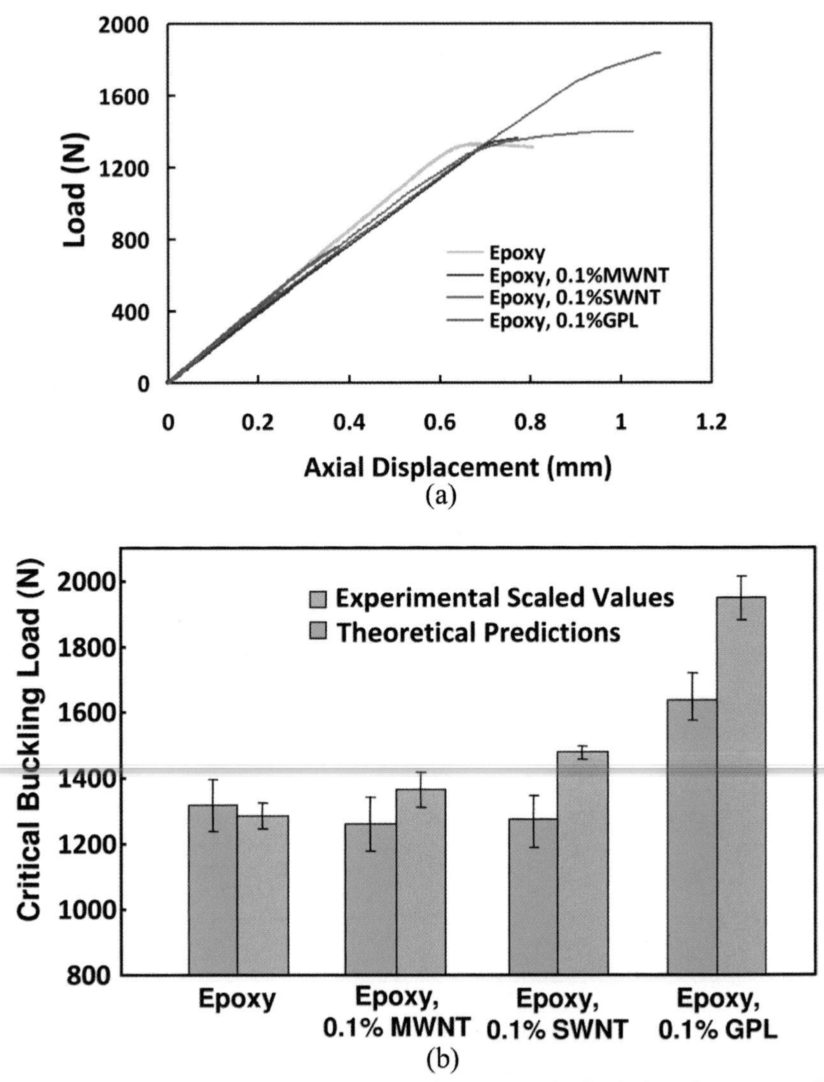

FIGURE 2.9. (a) Load-displacement curve under compression for the baseline epoxy and epoxy nanocomposites with 0.1% of GPL, MWNT, and SWNT additives, respectively. (b) Absolute values for the critical buckling load (1st mode) for the baseline epoxy and nanocomposite samples. (Adapted from [17] with permission).

the baseline epoxy specimen, the average scaled buckling load [Figure 2.9(b)] is ~1285 N, which is in reasonable agreement with the theoretical prediction (~1316 N) based on classical Euler buckling [Equation (20)] and the measured tensile modulus [Figure 2.8(b)]. Figure 2.9(b) indicates that, with the addition of ~0.1% weight MWNTs to the epoxy matrix, the average scaled buckling load is increased by ~6.2% to 1363 N. For 0.1% weight of SWNTs, the critical buckling load increases to 1477 N (a 15% increase). At the same nanofiller loading fraction of 0.1%, the GPL far outperforms the SWNT and MWNT additives and shows ~52% increase in the critical buckling load to 1947 N.

According to the classical Euler buckling equation, the buckling load is a function of geometry and elastic modulus of the material. Since the critical buckling load is corrected for geometry variation by proper scaling [Equation (23)], the increase in the buckling load should be accounted for by the increase in the elastic modulus of the nanocomposites compared to the pristine epoxy. Therefore, we expect the buckling loads to increase by ~0.5%, ~4%, and ~32% [based on the increase in the elastic modulus, Figure 2.8(b)], upon the addition of MWNTs, SWNTs, and GPLs, respectively. Nonetheless, the test data indicates ~6% (for MWNT), ~15% (for SWNTs), and ~52% (for GPLs) increases in the critical buckling load. This suggests that the elastic modulus of the carbon nanotube and graphene composites under large compressive loading (more specifically, at the onset of buckling) is significantly enhanced compared to tensile loading. Enhanced load transfer in carbon nanotube composites under compressive load has been reported by Schadler and co-workers [19]. They studied interfacial load transfer in nanotube epoxy composites under compressive loading by analyzing shifts of peaks in the Raman spectra. When the composite was subjected to a large compressive load, a larger shift in the Raman peak position was observed than in the tension case, which indicates that load transfer in nanotube bundles is improved under large compressive stresses. When the specimen is subjected to a tensile load, possibly only the peripheral surface of the nanotube bundles that are bonded to the epoxy matrix are effective for load transfer, and the weak intertube bonding contributes little to the load transfer. In contrast, when the specimen is subjected to compression, high compressive stress might force the polymer to infiltrate the nanotube bundles, creating a larger interaction zone between the polymer and the nanotube bundles, and thus improving the polymer–nanotube load transfer.

What is interesting is that, similar to nanotubes, an enhancement in the load transfer effectiveness under compressive stress is also observed for graphene. High resolution transmission electron microscopy characterization (Chapter 1) indicates that typical GPLs are comprised of ~3–4 individual graphene sheets. Similar to nanotube bundles, only the outer graphene sheets are expected to contribute to load transfer under tension while, under compression, the load may be shared more equitably by the sheets. Another consideration is that individual graphene sheets within clusters are expected to buckle and bend under compressive stress due to their atomic scale thickness. This bowing (or bending) of individual graphene sheets within clusters can increase frictional interlocking between sheets and, thus, provide better sheet-to-sheet load transfer within the bundle. To summarize, graphene composites show potential to provide significant enhancement in buckling stability, which is an important consideration for the design of ultra lightweight and highly optimized structural elements used in aeronautical and space applications.

2.4. FRACTURE TOUGHNESS

The discrepancy between the theoretical (ideal) strength and the experimentally observed fracture strength of most engineering materials is explained by the presence of inherent flaws (defects) at the material surfaces or in the interior. It is essential to quantify the material toughness in the presence of such flaws, which falls under the realm of fracture mechanics. Fracture toughness quantifies the ability of a material containing a crack (defect) to resist fracture and it is a critically important material property for design applications. Defects such as microcracks, voids, or delaminations are inevitable during the manufacturing of structural components. It is, therefore, highly desirable to have materials that resist crack growth originating from such defect sites. Materials with high fracture toughness offer enhanced safety, reliability, and operating life of structural components.

Linear elastic fracture mechanics [20] has been vital in understanding materials and fracture at the microscopic level. Under a given applied stress, there are three basic crack opening modes: Mode I in normal tension, Mode II in-plane shear, and Mode III out-of-plane shear, as shown in Figure 2.10.

For the Mode I configuration, the stresses acting on any element of the material are obtained using principles of elastic theory. The tensile and shear stresses in an element at a distance r and angle θ from the crack tip are given as:

$$\sigma_x = \frac{k}{\sqrt{2\pi r}} f_x(\theta) \tag{24}$$

$$\sigma_y = \frac{k}{\sqrt{2\pi r}} f_y(\theta) \tag{25}$$

$$\tau_{xy} = \frac{k}{\sqrt{2\pi r}} f_{xy}(\theta) \tag{26}$$

where k is the stress intensity factor, a function of applied stress, σ, and crack length, a, and is given by:

$$k = Y\sigma\sqrt{\pi a} \tag{27}$$

where Y is a dimensionless parameter governed by sample and crack geometry.

Since the stress intensities in the crack vicinity can be described in terms of k, its critical value is defined to specify the condition for brittle failure. It is given at a critical stress (σ_o) corresponding to a critical crack size, a_c. This critical k value is called the fracture toughness (K_c) of the material, given by:

$$K_c = Y\sigma_{oc}\sqrt{\pi \times a_c} \tag{28}$$

Its value is reported under plain strain conditions (sample with thickness $\geq 2.5 \times [K_{IC})/(\sigma_y)]^2$, where σ_y is the yield stress) for Mode I loading, indicated as K_{IC}. K_{IC}, also called plain strain fracture toughness, is defined as the ability of a material containing a crack to resist failure when stress is applied in Mode I. K_{IC} is related to the critical energy

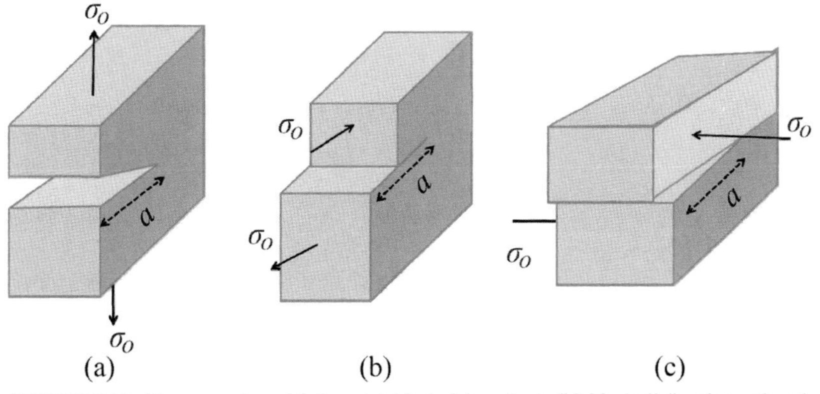

FIGURE 2.10. *Three modes of failure (a) Mode I (tension), (b) Mode II (in-plane shear), and (c) Mode III (out-of-plane shear).*

release rate G_{IC} in Mode I, defined as the amount of energy released per unit area as the crack propagates, given by:

$$G_{Ic} = \frac{K_{IC}^2(1-v^2)}{E} \tag{29}$$

where v is the Poisson's ratio of the material. The strain energy release rate governs the crack propagation rate.

A compact tension sample geometry, depicted in Figure 2.11, is used for measuring fracture toughness K_{IC} for plastics (ASTM Standard D5045). An initial pre-crack is created in the compact tension samples by gently tapping a fresh razor blade over a molded starter notch. Monotonic fracture toughness tests can be performed on polymer composites using the ASTM Standard D5045, applying the equation:

$$K_{Ic} = \frac{P_{max}}{BW^{1/2}} f\left(\frac{a}{W}\right) \tag{30}$$

where P_{max} is the maximum load in the load-displacement curve for the compact tension specimen, B is the specimen thickness, W is the specimen width, and $f(a/W)$ is related to the sample geometry, given by:

$$f\left(\frac{a}{W}\right) = \frac{\left[\left(2+\frac{a}{W}\right)\left(0.886+4.64\left(\frac{a}{W}\right)-13.32\left(\frac{a}{W}\right)^2+14.72\left(\frac{a}{W}\right)^3-5.6\left(\frac{a}{W}\right)^4\right)\right]}{\left(1-\frac{a}{W}\right)^{3/2}} \tag{31}$$

A typical sample geometry is shown schematically in Figure 2.11 and its dimensions are provided in Table 2.1.

In this section, crack opening tests on compact tension samples [16] are described to measure the Mode I fracture toughness (K_{Ic}) of the pure epoxy matrix and the GPL-epoxy nanocomposites at various GPL weight fractions. At each weight fraction of GPL additives, between 4–6 different samples were tested to check for reproducibility of the results. Figure 2.12 shows that the Mode I fracture toughness (K_{Ic}) of the baseline epoxy (without GPL) is ~1.03 MPa $\sqrt{m}$, which correlates

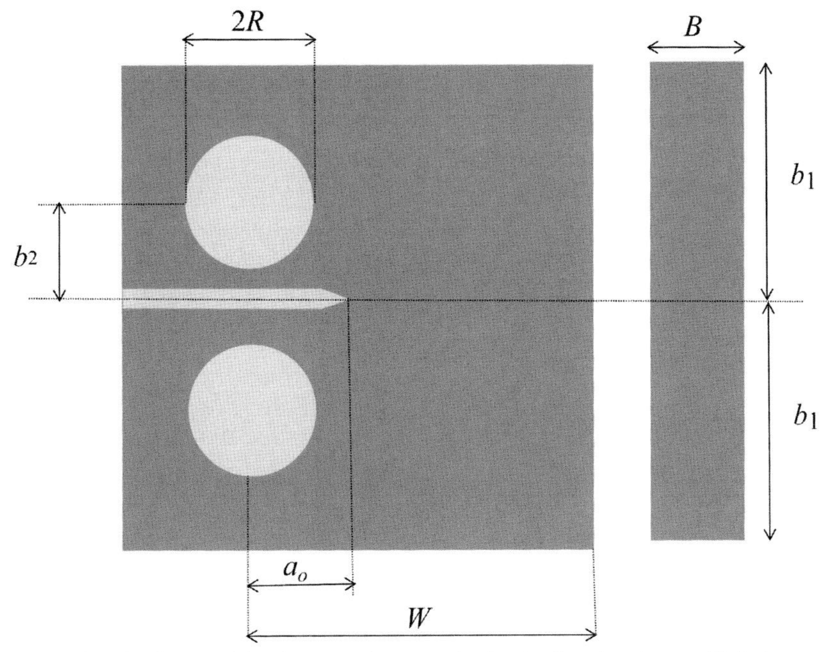

FIGURE 2.11. Compact tension sample geometry for fracture toughness (K_{IC}) characterization.

well with published literature for epoxy materials. Addition of GPL into the epoxy matrix causes a sharp increase in the nanocomposite's K_{Ic} to ~1.75 MPa $\sqrt{m}$ at 0.125% weight fraction of GPL, which corresponds to a ~65% increase in fracture toughness. For higher loading fractions, the enhancement in K_{Ic} diminishes (Figure 2.12) and finally begins to approach the pure epoxy value at a GPL weight fraction of ~0.5%. This might be a result of degradation in the dispersion of the

TABLE 2.1. The Sample Geometry as Prescribed for K_{IC} Test.

Symbol	Description	Dimension (mm)
W	Width	25.4
L	Overall length	31.75
$2 \times b_1$	Full breadth	30.5
$2 \times b_2$	Distance between the centers of the holes and the crack plane	13.97
R	Radius	3.2
B	Thickness	12.7
a_o	Initial crack length	5

GPL additives above a weight fraction of 0.125%. Similar degradation in the dispersion quality of carbon nanotubes in epoxy composites has been observed [21–22], but at significantly larger nanotube weight fractions (> 0.5%). Therefore, dispersion of 2D GPL appears to be more challenging as compared to 1D carbon nanotubes or $Al_2O_3/TiO_2/SiO_2$ nanoparticles [23–25], which have been successfully dispersed in epoxies at very high loading fractions (up to 20%). This is not an unexpected result given that the 2D GPL (order of several microns in dimension) are more easily entangled and, therefore, show a greater tendency to agglomerate compared to nanoparticles or nanotubes. In spite of this, the maximum enhancement in K_{Ic} (~65% at 0.125% weight fraction of GPL) is impressive. To achieve comparable increase (~62%) in K_{Ic}, the required weight fraction (~14.8%) of SiO_2 nanoparticles [23] in epoxy is ~120-fold larger than GPL. Similarly, to obtain a 65% increase in K_{Ic}, the volume fraction of Al_2O_3 (~5%) and TiO_2 (~10%) nanoparticles in epoxy is ~30 to 60-fold larger than GPL [24–25].

Intercalated nanoclays [26–28] have also proven to be effective in toughening epoxy systems. For example, Zerda *et al.* obtained ~61% increase in fracture toughness for 3.5% nanoclay weight fraction. Wang *et al.* obtained similar levels of K_{Ic} enhancement (~78%) at 2.5% weight

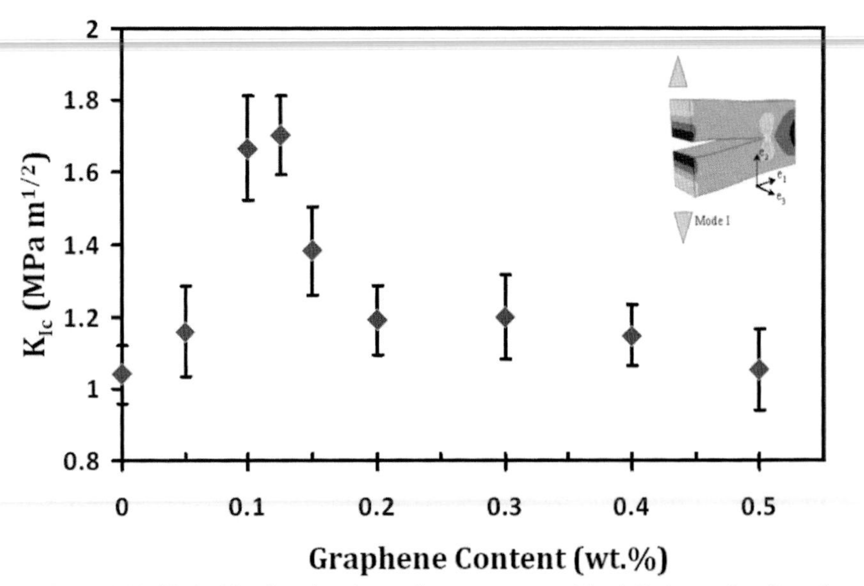

FIGURE 2.12. Mode I fracture toughness for an epoxy matrix plotted as a function of the graphene content. Inset shows a schematic of the Mode I crack opening test used to determine the fracture toughness. (Adapted from [16] with permission).

fraction of nanoclay fillers. Liu *et al.* demonstrated ~110% increase in K_{Ic} for an epoxy system, with ~3% weight of nanoclay additives. These weight fractions are 20–30 times higher than the optimal GPL weight fraction of 0.125%. Moreover, in the case of nanoclay epoxy composites, addition of nanoclays (in 2.5–3.5% weight fraction range) causes degradation in the tensile strength. For example, the tensile strength of the matrix [27] was reduced by ~17% due to incorporation of ~2.5% weight of nanoclay. Similarly, the tensile strength decreased [26] by ~7.5% for ~3.5% weight fraction of nanoclays in the epoxy matrix. By contrast, the GPL additives were found to enhance the tensile strength by up to ~45% in the ~0.1–0.125% weight fraction range [Figure 2.7(a)].

For carbon nanotube epoxy composites, the best reported enhancement in K_{Ic} is ~43% [29], which occurs at ~4-fold higher nanofiller weight fraction (~0.5% weight of amine-functionalized double-walled carbon nanotubes). In multi-walled carbon nanotube epoxy composites, only ~11% improvement in fracture toughness [21] can be achieved with ~0.25% weight fraction of multi-walled carbon nanotube additives. These results indicate that GPL are highly effective in suppressing crack propagation in polymer materials. However, there is a clear need to develop improved techniques for GPL dispersion in polymer matrices in order to realize their full potential. For example, pre-modifying GPL by use of specially designed surfactants or functionalization with amine groups may enable enhanced dispersion at higher GPL loading fractions (> 0.125%).

The measurement of modulus (E) and fracture toughness (K_{Ic}) enables us to compute the critical energy release rate: $G_{Ic} = K_{Ic}^2[(1-\mu^2)/E]$, where μ is the Poisson's ratio. The fracture energy (G_{Ic}) quantifies the energy required to propagate the crack in the material. Figure 2.13 indicates that the G_{Ic} of the baseline epoxy (without GPL) is ~ 325 J/m^2, which correlates well with published literature for brittle polymers [16], which typically show G_{Ic} less than 500 J/m^2. Incorporation of GPL into the epoxy causes a large increase in the nanocomposite's G_{Ic} to ~700 J/m^2 at 0.125% weight fraction of GPL, which corresponds to an ~115% increase. The critical energy release rate for the GPL/epoxy nanocomposite is comparable to brittle metals [16], which typically display G_{Ic} in the range of 800–2000 J/m^2.

It is also interesting to compare the fracture toughness and fracture energy of GPL/epoxy composites to that of SWNT and MWNT epoxy composites at the same nanofiller loading fraction of ~0.1%. Figure 2.14 [18] indicates that, compared to the epoxy, the SWNT, MWNT,

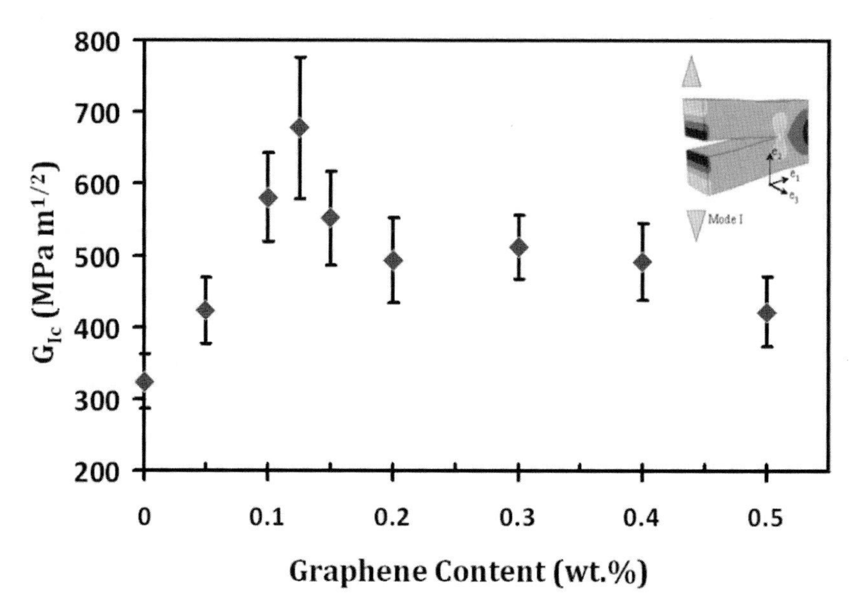

FIGURE 2.13. *Mode I fracture energy for an epoxy matrix plotted as a function of the graphene content. Inset shows a schematic of the Mode I crack opening test. (Adapted from [16] with permission).*

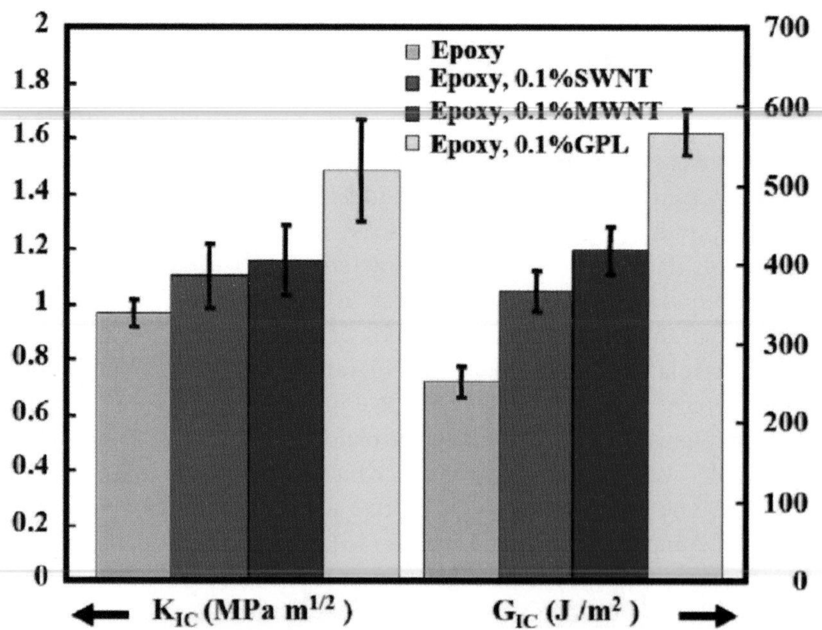

FIGURE 2.14. *Mode I fracture toughness and Mode I fracture energy for baseline epoxy and nanocomposites with various types of nanofillers. The nanofiller loading fraction is fixed at ~0.1% to compare the effectiveness of the various types of additives. (Adapted from [18] with permission).*

and GPL nanocomposites show ~14%, ~20%, and ~53% increases, respectively, in fracture toughness (K_{Ic}). For the fracture energy (G_{Ic}), ~45%, ~66%, and ~126% enhancement is observed for the SWNT, MWNT, and GPL nanocomposites, respectively. The ability of the GPL in toughening the matrix is clearly superior to carbon nanotubes. This is also corroborated by the work of other groups. For example, previous studies [29] have reported increases of only 18% and 26% in the fracture toughness of epoxy composites with 0.1 and 1 wt. %, respectively, of amine-functionalized double-walled carbon nanotubes. The fracture toughness of GPL is also impressive in relation to nanoparticle composites. For example, the weight fraction of SiO_2 nanoparticles (7.8%) required [23–28] to increase the fracture toughness of the neat epoxy by 54% is *~80-fold higher* compared to GPL. Similarly, clay nanocomposites [26–28] require between 5 and 10 weight % of nanoclay additives in various epoxy resins to achieve a 60% increase in the fracture toughness, which is *50- to 100-fold larger* than the GPL weight fraction of 0.1%.

2.5. FATIGUE RESISTANCE

Fatigue involves dynamic propagation of cracks under cyclic loading and it is one of the primary causes for catastrophic failure in structural materials. Consequently, the material's resistance to fatigue is of paramount importance to prevent failure. Fatigue resistance is the resistance offered by the material to a subcritical crack that propagates under alternating stresses lower than the material's ultimate tensile stress. Fatigue crack propagation rate (FCPR) and fatigue life measurements based on fracture mechanics models are the two ways to characterize the fatigue resistance or the fatigue life of the material [30–32].

Fatigue crack propagation rate: The FCPR of a material is divided into three stages, as shown in Figure 2.15(a). In Stage I, the crack growth is below the threshold value of the stress intensity factor. The threshold stress intensity factor (Δk_{th}) is defined at the stress range below which cracks will not propagate in the material. In the fatigue life diagram, it is indicated by the fatigue limit. In Stage II, the size of a crack grows steadily, and its growth is governed by the Paris law given by:

$$\frac{da}{dN} = C.(\Delta k)^n \tag{32}$$

The Paris law states that the crack propagation rate per cycle (*da*/*dN*) is directly related to the stress intensity factor range;

$$\Delta k = (k_{max} - k_{min}) = Y(\sigma_{max} - \sigma_{min})(\pi a)^{1/2} \tag{33}$$

And, on external testing conditions, described by two constants C and n. The higher the crack growth rate, the shorter the fatigue life. Stage III is the terminal regime, where the crack growth is unstable and the material finally fails.

Fatigue life diagram: The fatigue life diagrams or *S*-*N* (stress-number of cycles) curves are used to estimate the expected number of life cycles (N_f) a material can withstand under a given stress (S). At high S values, N_f is small and the region is called low cycle fatigue, while at low S values, N_f is large, corresponding to the high cycle fatigue region. The *S*-*N* curve of a typical material is shown in Figure 2.15(b). The fatigue life of a material is especially difficult to predict as cracks propagate under mixed loading. Ductile materials commonly show both Modes I and II of crack propagation. The initial propagation of fatigue cracks in materials has been observed to be under shear mode while, in the later stages of crack propagation, it propagates under tensile mode. Typically compact tension based geometries based on ASTM E647 and taper double cantilever beam (DCB) geometry are used for FCP testing.

In this section, fatigue crack propagation tests on compact tension samples following the ASTM E647 standard are presented. Figure 2.16 shows the measured crack propagation rate (*da*/*dN*) versus the applied stress intensity factor amplitude (ΔK). A substantial lowering in the crack growth rate over the full range of stress intensity factor amplitudes can be observed for the GPL nanocomposite [18] compared to the baseline epoxy. For example, at $\Delta K = 0.5$ MPa m$^{1/2}$, the *da*/*dN* for the nanocomposite (5.87×10^{-5} mm/cycle) is ~40-fold lower than the baseline epoxy (2.5×10^{-3} mm/cycle). The crack growth results can be fitted to the Paris-Erdogan law [Equation (32)]. A significant reduction in the exponent n (from ~17.33 to ~7.94) and the constant C (from ~394.26 to ~0.0358) for the graphene nanocomposite compared to the baseline epoxy can be observed.

Figure 2.16 also compares [18] the fatigue suppression performance of GPL with SWNT and MWNT additives at the same nanofiller weight fraction of ~0.1%. The performance of the GPL is clearly superior to the nanotubes, particularly as the stress intensity factor amplitude (ΔK) is increased. For the case of the nanotubes, one can observe a substantial degradation in the fatigue suppression with increasing ΔK (Figure 2.16). This is because, in the case of nanotubes, the dominant toughening and fatigue suppression mechanism is crack bridging. Wei and co-workers have shown that the fatigue crack is bridged by high aspect ratio nanotubes [33–34] generating a fiber-bridging zone in the wake of the crack tip. As the crack advances, energy is dissipated by the frictional pull-out of the bridging nanotubes from the epoxy matrix, which

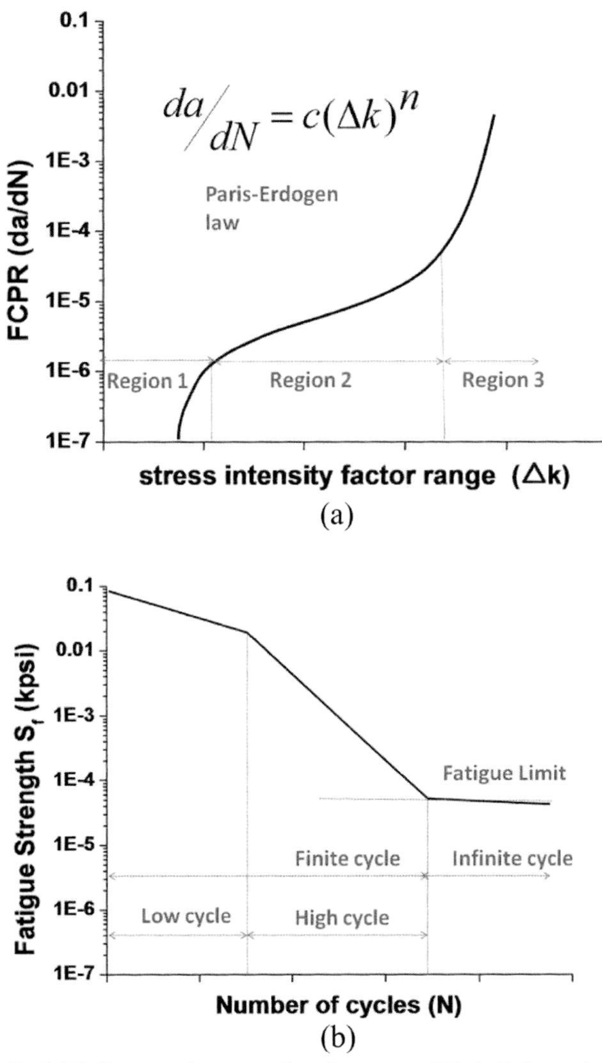

FIGURE 2.15. (a) Fatigue crack propagation diagram and Paris-Erdogen law (b) S-N or fatigue life diagram.

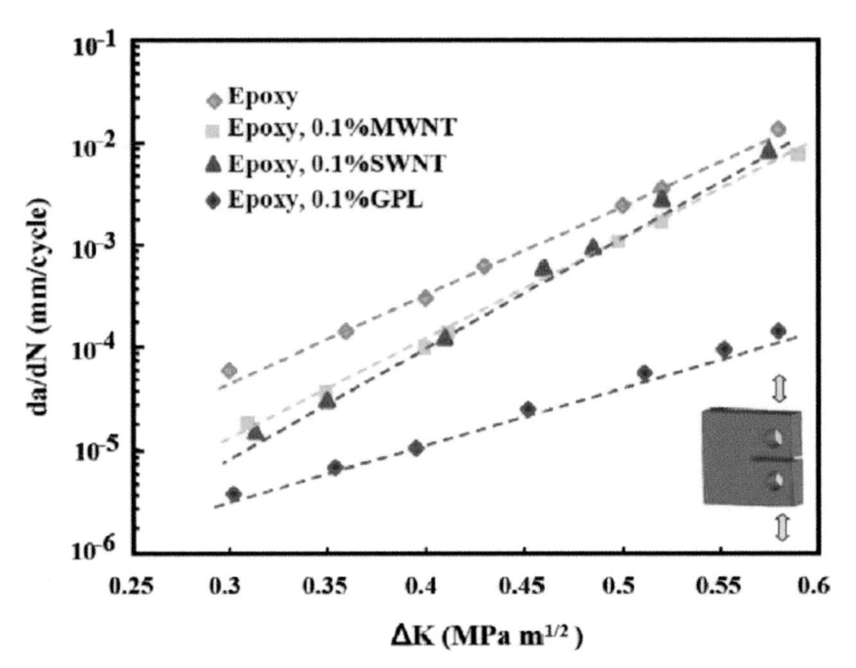

FIGURE 2.16. Fatigue crack propagation testing; crack growth rate (da/dN) plotted as a function of the stress intensity factor amplitude (ΔK) for the pristine epoxy and nanocomposite samples with ~0.1 wt % of GPL, ~0.1 wt % of SWNT, and ~0.1 wt % of MWNT additives. (Adapted from [18] with permission).

slows the crack propagation speed. However, this crack bridging effect loses effectiveness at high ΔK due to progressive shrinkage in the size of the fiber bridging zone as ΔK is increased. The fact that such behavior is not observed in GPL/epoxy nanocomposites indicates that the toughening mechanism for GPL is different from nanotubes. Unlike nanotubes, frictional pull-out of GPL from the matrix is less likely given the enhanced interfacial adhesion of GPL with polymer matrices.

2.6. TOUGHENING MECHANISMS

A variety of mechanisms could contribute to enhancing the fracture toughness and fatigue resistance of polymers by the graphene fillers. Some of the key mechanisms are summarized below:

Crack deflection: Crack deflection [16,35] at nanofiller interfaces could be an important source of increased energy dissipation for graphene nanocomposites. When a crack front encounters a second phase material, it either cuts them or passes around them; i.e., it deflects from the original path. Propagation of cracks through brittle fillers does not improve the polymer toughness, as the energy dissipation in brittle failure is insignificant. But, if cracks tilt and twist around the nanofillers, it increases the total fracture surface area by redirecting the crack. A way to verify crack deflection as the toughness mechanism is by correlating the fracture toughness values with the fracture surface roughness. The deflection of the crack around the filler and, hence, the increased fracture toughness is a function of the matrix strength, filler strength, and the nature of the interface between them. During crack deflection, tilts and twists of the crack front occur around the nanofillers, which cause the crack front to switch the mode of propagation between I/II (tilt) and I/III (twist), as shown in Figure 2.17, for platelet nanofillers. The crack deflection toughening mechanism has been treated well by Faber and Evans in their classic work in 1982, the mathematics of which are detailed below.

Crack Deflection Theory: Faber and Evans [35] analyzed the mechanism of crack deflection in detail for different filler geometries. On encountering second phase fillers, the local stress intensity at the crack tip changes. The stress intensity factors for tilt (k_1^t, k_2^t) and twist (k_1^T, k_3^T) were derived by them to be:

$$k_1^t = k_I \cos^3(\theta/2) \tag{34}$$

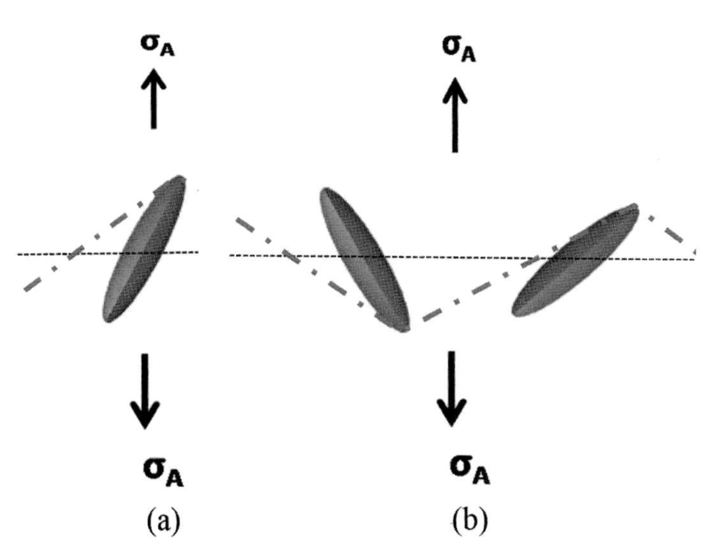

FIGURE 2.17. *Schematic of crack front undergoing (a) tilt and (b) twist under mixed mode on encountering second phase particles.*

$$k_2^t = k_I \sin(\theta/2)\cos^2(\theta/2) \tag{35}$$

$$k_3^T = \begin{pmatrix} \cos^4(\theta/2)[\sin\psi\cos\psi(\cos^2(\theta/2) - 2\upsilon)]\}\cos^2\psi]\}k_1^t + \\ \{\sin^2(\theta/2)\cos^2(\theta/2)X[\sin\psi\cos\psi(3\cos^2(\theta/2) - 2\upsilon]\}k_2^t \end{pmatrix} \tag{36}$$

$$k_1^T = \begin{pmatrix} \{\cos^4(\theta/2)[2\upsilon\sin^2\psi + \cos^2(\theta/2)\cos^2\psi]\}k_1^t + \\ \{\sin^2(\theta/2)\cos^2(\theta/2)X[2\upsilon\sin^2\psi + 3\cos^2(\theta/2)\cos^2\psi]\}k_2^t \end{pmatrix} \tag{37}$$

where k_1^t and k_2^t are the local stress intensity factors for the tilted crack in Modes I and II, respectively, and k_1^T and k_3^T are the local stress intensity factors for the twisted crack under Modes I and III, respectively; k_I is the stress intensity factor of the material in Mode I, θ is the tilt angle, ψ is the twist angle and υ is the Poisson's ratio. Then, the strain energy release rate corresponding to each misaligned section of the crack can be given by:

$$G = \{k_1^2(1-\upsilon^2) + k_2^2(1-\upsilon^2) + k_3^2(1-\upsilon^2)\}/E \tag{38}$$

The nearest neighbor distances for the platelets were calculated using an extension of the finite parallel cylinder model, where:

$$\Delta_{cyl} \Big/ r_{cyl} = \{e^{4Vf} \Big/ V_f^{1/2}\} \int_{4Vf}^{\infty} x^{1/2} e^{-x} dx \tag{39}$$

where Δ_{cyl} and r_{cyl} are the intercylinder spacing and radius of finite cylinders and V_f is filler volume fraction.

Using Equation 39, they obtained the inter-filler spacing to be:

$$\langle\Delta\rangle_{disc} \Big/ \bar{t} = R^{1/3} * \Delta_{cyl} / r_{cyl} \tag{40}$$

where $R = (H/t)$ is the aspect ratio of the platelet, with H being the average platelet length and t its average thickness. The net toughness enhancement was obtained using the Equations (34)–(40) by Faber and Evans and the expressions are given in Equations (41) and (42), for twist and tilt, respectively:

$$\langle G\rangle^T / G_o =$$

$$\frac{4}{\pi^4} \int_{-\pi/2}^{\pi/2} \int_{-\pi/2}^{\pi/2} \int_0^1 \int_0^1 \int_{-\pi/2}^{0} \int_0^{\pi/2} \eta \left[\begin{array}{c} \cos^4(\langle\theta\rangle/2) \left\{ \begin{array}{c} 2\upsilon \sin^2\psi + \cos^2(\langle\theta\rangle/2)\cos^2\psi \\ *[1+\sin^2(\langle\theta\rangle/2)] \end{array} \right\}^2 \\ +\{\sin\psi \cos\psi[\cos^2(\langle\theta\rangle/2) - 2\upsilon \end{array} \right] +$$

$$\sin^2(\langle\theta\rangle/2)[3\cos^2(\langle\theta\rangle/2) - 2\upsilon]\}^2] d\theta_1 d\theta_2 d\alpha d\beta d\mu_1 d\mu_2$$

(41)

$$\langle G\rangle^t / G_o = \frac{4}{\pi^4} \int_{-\pi/2}^{\pi/2} \int_{-\pi/2}^{\pi/2} \int_0^1 \int_0^1 \int_0^{\pi/2} \int_0^{\pi/2} \xi \cos^4(\bar{\theta}/2) d\theta_1 d\theta_2 d\alpha d\beta d\mu_1 d\mu_2 \quad (42)$$

where $G_o = k_I(1 - \upsilon^2)/E$ is the strain release rate for an undeflected crack, and η and ξ are the ratios of undeflected to deflected crack length for twist and tilt, respectively, and are given by:

$$\eta = \frac{[(\Delta/H) - \alpha\cos\theta_1 \sin\mu_1 + (1-\beta)\cos\theta_2 \sin\mu_2]}{\{(\Delta^*)^2 + [\alpha\sin\theta_1 + (1-\beta)\sin\theta_2]^2\}^{1/2}} \quad (43)$$

$$\xi = \frac{[(\Delta/H) - \alpha\cos\theta_1 \sin\mu_1 + (1-\beta)\cos\theta_2 \sin\mu_2]}{\{(\Delta^*)^2 + [\alpha\sin\theta_1 - (1-\beta)\sin\theta_2]^2\}^{1/2}} \quad (44)$$

where α and β are the relative orientations where the fracture plane cuts two nearest neighbor platelets, $<\theta>$ is the average tilt angle in the twist plane, $\bar{\theta}$ is the average tilt angle, and ψ is the twist angle and is given by:

$$\Psi = \tan^{-1}\{[\alpha\sin\theta_1 + (1-\beta)\sin\theta_2]/\Delta^*\} \quad (45)$$

where Δ* is given by:

$$\Delta^* = \left\{ \begin{array}{c} [(\Delta/H) - \alpha\cos\theta_1 \sin\mu_1 + (1-\beta)\cos\theta_2 \sin\mu_2]^2 + \\ [\alpha\cos\theta_1 \cos\mu_1 - (1-\beta)\cos\theta_2 \cos\mu_2]^2\}^{1/2} \end{array} \right\} \quad (46)$$

The total strain energy release rate is a combination of the strain energy rate due to crack tilt and twist given as:

$$\langle G\rangle_{rod} = \eta/2\langle G\rangle^{T} + \xi/2\langle G\rangle^{t} \tag{47}$$

Some of the assumptions that have been made for the above derivations include constant interparticle spacing and uniform platelet size. Crack front twisting has been found to be the major toughening mechanism for spherical and rod-shaped fillers. In contrast, for disc-shaped fillers, the initial tilt of the crack front was found to play the most significant role in toughening. The approximate contributions to toughening from the initial crack tilt for spherical, rod, and disc fillers were calculated [35] to be:

$$\frac{(G_C^t)Sphere}{(G_C^{Matrix})} = 1 + 0.87V_f \tag{48}$$

$$\frac{(G_C^t)Rod}{(G_C^{Matrix})} = 1 + V_f(0.6 + 0.014(h/2r) - 0.0004(h/2r)^2) \tag{49}$$

$$\frac{(G_C^t)Plate}{(G_C^{Matrix})} = 1 + 0.28V_f(l/t) \tag{50}$$

where

$$(G_C^t)Sphere/(G_C^{Matrix}),\ (G_C^t)Rod/(G_C^{Matrix}) \text{ and } (G_C^t)Plate/(G_C^{Matrix})$$

are the fracture energy ratio of nanocomposite and pristine matrix for spherical-, tubular- and plate-shaped fillers. V_f is the filler content in volume fraction, ($h/2r$) is the rod length (h) to its radius (r), and (l/t) is the plate's diameter (l), to its thickness (t).

The toughening enhancement due to the initial tilt is plotted as a function of the aspect ratio for various geometries of fillers and is shown in Figure 2.18(a). It is seen that disc- (platelet) shaped fillers obtain maximum toughening increment due to initial crack tilt, especially when their aspect ratio is high. The distinguishing feature of graphene compared to other forms of 2D fillers is its very high aspect ratio. However, in reality the aspect ratio of the graphene disk will be much lower than the ideal theoretical value due to the curvature and waviness of the flexible graphene sheets in the polymer matrix. This is similar to a flexible polymer chain that displays a characteristic radius of gyration.

Therefore, the effective aspect ratio of the graphene sheet is expected to be at least an order of magnitude smaller than the ideal value. Assuming an average GPL diameter of ~2 microns and thickness of ~2 nm, the ideal aspect ratio equates to ~1000. In Figure 2.18(b), this aspect ratio is reduced to 100 to account for curvature effects. For this reduced aspect ratio, the Faber/Evans model [35] shows excellent correlation with test data until about ~0.125% weight fraction, after which the test data rapidly degrades due to deterioration in the GPL dispersion.

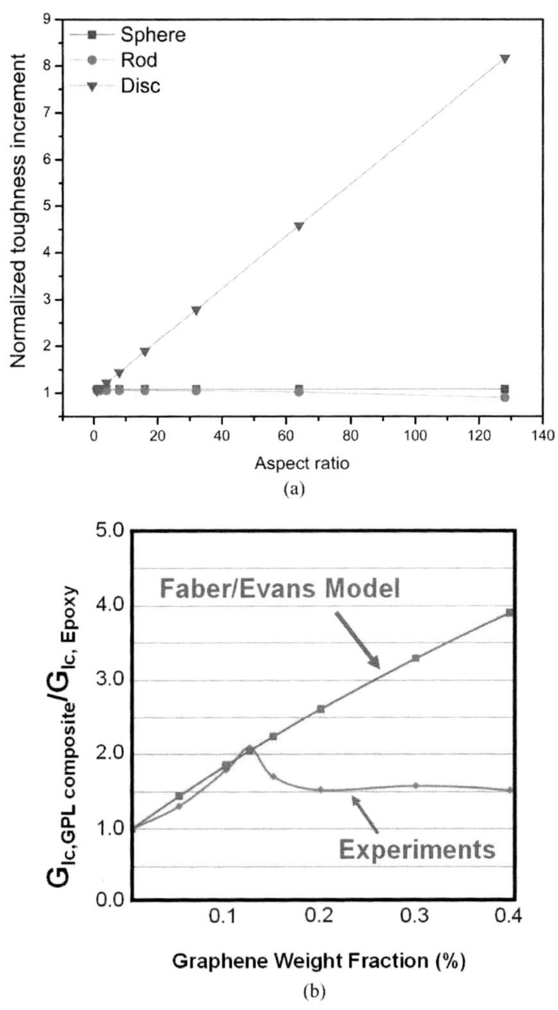

FIGURE 2.18. *(a) Normalized toughening increment for disc, spherical, and rod fillers for initial crack tilt for volume fraction 0.1, calculated from Equations (48)–(50). (b) Equation (50) for a disk filler with aspect ratio of ~100 compared with test data.*

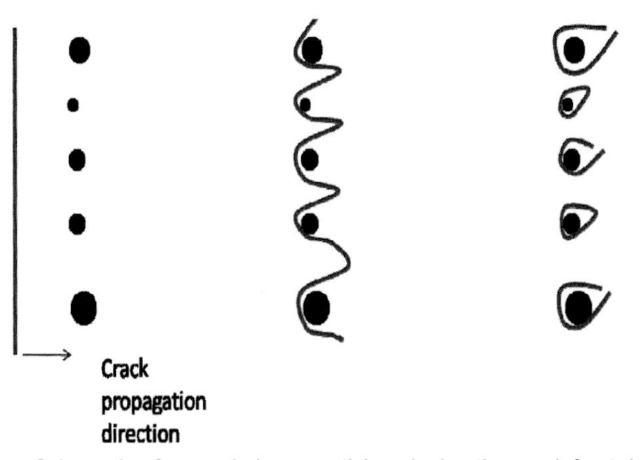

FIGURE 2.19. Schematic of second phase particles pinning the crack front, leaving behind tail-like structures. (Adapted from [36] with permission).

Crack Pinning: Crack pinning as an efficient toughening mechanism was proposed by Lange [36] for arresting advancing cracks in composites with a second phase at an appropriate interparticle spacing. It is very commonly observed in inorganic nanoparticle-reinforced epoxy nanocomposites. Crack pinning increases the amount of energy dissipated by the composite during fracture by forcing the crack front to bow in between the nanofillers. A crack possesses a line tension, and bowing between fillers increases the crack length and, hence, the energy absorbed for propagation. The pinned cracks likely create new surfaces and nucleate secondary cracks, which form a tail-like structure, as shown in Figure 2.19, leading to enhanced energy absorption.

Debonding and Plastic Void Growth: Debonding or pull out of nanofillers from the polymer matrix (Figure 2.20), followed by plastic void growth of the polymer, is a vital toughening mechanism when filler-matrix bonding is sufficiently strong. The particle-debonding does not dissipate as much energy as the energy absorption by the epoxy matrix during its plastic deformation after filler debonding. The debonding energy is a strong function of particle-matrix adhesive forces; strongly adherent fillers debond at a certain stress, reducing the stress constraint at the crack tip and allowing the matrix to deform [37].

Fiber Bridging and Pull-out: Bridging of the crack front by high aspect ratio fillers and their pull out from the matrix [34–35] is a toughening mechanism observed in rod-shaped fillers reinforcing polymer nanocomposites. The bridging fillers lower the stress concentration at the crack tip, as shown in Figure 2.21.

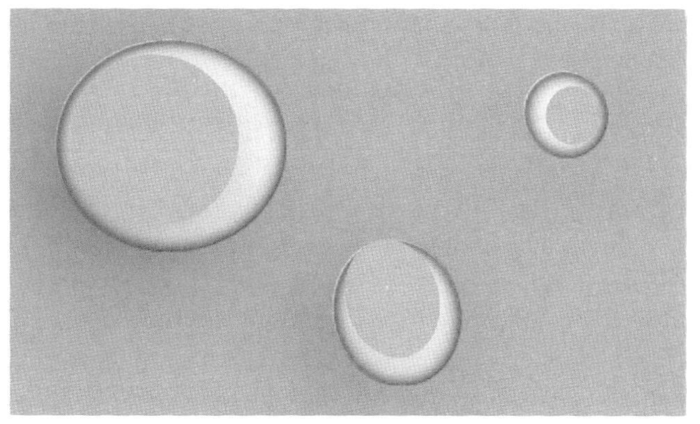

FIGURE 2.20. *Schematic of particles debonded from matrix.*

Microcracking: Microcracking [38] has been observed as an effective toughening mechanism in polymer nanocomposites. Microcracks can form in the matrix through the filler, between the layers of platelet fillers, and at the filler-matrix interface. Voids result from opening of these microcracks, relieving the constraint and permitting extra strain for the matrix. Similar to debonding of the filler, microcracking allows plastic void growth of the matrix.

Plastic-zone Branching: Plastic zone branching is a synergistic mechanism reported for hybrid composites of inorganic and elastomer fillers for polymer toughening. The high stress regions created in the

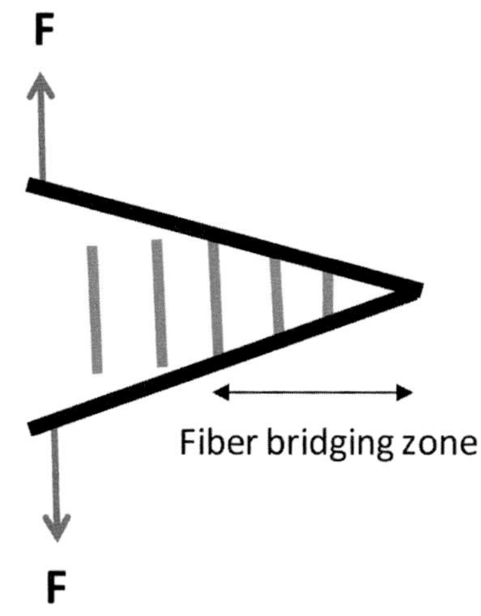

FIGURE 2.21. *Schematic of crack bridging and pull out of high aspect ratio fillers in the crack wake.*

matrix due to the inorganic particles cause cavitation and shear banding of and around the elastomers, in addition to shear deformation of the matrix between the inorganic fillers [39].

Crack Tip Blunting: While not an exclusive mechanism by itself, crack tip blunting dictates the crack propagation rate. Mechanisms like localized plastic shear, debonding, cavitations, microcracking, and fracture of fillers help in blunting the crack tip [40].

To analyze which of the above mechanisms is playing an important role, fractography analysis of the fracture surface of the compact tension samples is typically performed. In the case of graphene composites, scanning electron microscopy (SEM) analysis of the fracture surface does not reveal any direct evidence of crack pinning or crack bridging by the GPL additives. In contrast, for carbon nanotube composites, crack bridging is a commonly observed toughening mechanism. For graphene composites, the most interesting observation from the fracture surface is a significant increase in average surface roughness of the fracture surface with GPL content [16], as indicated in Figure 2.22(a). The data for the average surface roughness (R_a) of the fracture surfaces (measured using a Dektak Surface Profiler from VEECO) indicates a doubling in the average surface roughness with increase in GPL content from 0 to 0.125% weight. This roughening effect begins to saturate with further increase in the GPL content. The increasing roughness of the fracture surface with GPL content suggests that crack deflection appears to play a significant role in the observed toughening. Crack deflection is the process by which an initial crack tilts and twists when it encounters a rigid inclusion. This generates an increase in the total fracture surface area resulting in greater energy absorption as compared to the unfilled polymer material. The tilting and twisting of the crack front as it is forced to move out of the initial propagation plane also forces the crack to grow locally under mixed-mode (tensile/in-plane shear and tensile/anti-plane shear) conditions. Crack propagation under mixed-mode conditions requires a higher driving force [16] than in Mode I (tension), which also results in higher fracture toughness of the material. If such crack deflection processes are playing a major role, then we expect a linear increase in the fracture surface roughness (R_a) as the fracture energy (G_{Ic}) is increased. This indeed appears to be the case in the 0–0.125% GPL weight fraction range as indicated in Figure 2.22(b). However, for higher weight fractions, the trend reverses, which suggests that a competing mechanism (probably agglomeration of GPL) that decreases the fracture toughness comes into play.

The classical Faber and Evans model provides insight into why crack deflection is more effective for high aspect ratio GPL sheets as compared to conventional fillers. Figure 2.18 shows the normalized toughening increment generated by crack tilting vs. the particle aspect ratio (as predicted by the Faber and Evans model) for particles with rod, sphere, and plate geometries. The results indicate that, for plates with

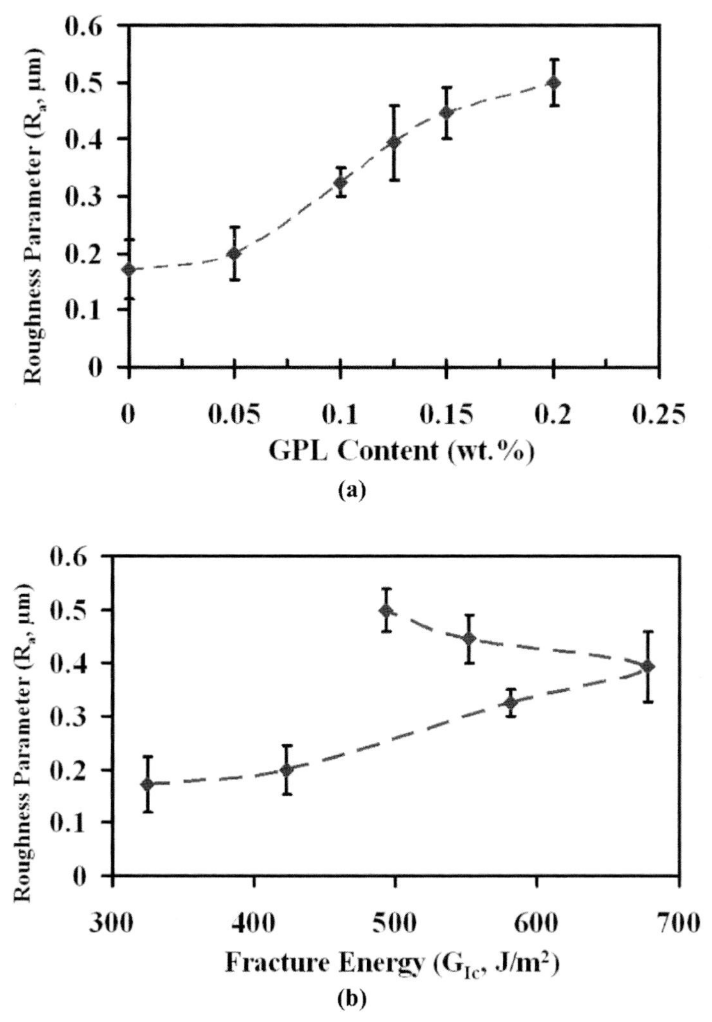

FIGURE 2.22. Mechanism for toughening. (a) Fracture surface roughness parameter (R_a) plotted as a function of the weight fraction of GPL in the epoxy matrix. (b) Variation in R_a as a function of the fracture energy (G_{Ic}). The linear increase of R_a with G_{Ic} (in the 0–0.125% GPL weight fraction range) indicates that crack deflection processes appear to play a significant role in the toughening. (Adapted from [16] with permission).

large aspect ratio (such as a GPL), the tilting of the crack front acts as a significant source of toughening. It is also evident that neither the sphere- nor the rod-shaped particles derive noticeable toughening from the crack tilting process. Furthermore, the aspect ratio of the rod has little effect on the toughening during the crack tilt process. This indicates that the GPL sheet geometry, with its very large aspect ratio, is expected to be highly effective in toughening the matrix, which is consistent with the experimental observations.

2.7. CHARACTERIZING THE GRAPHENE/MATRIX INTERFACE

In any nanocomposite material, the strength of the interface formed between the nanofiller and the surrounding matrix plays a crucial role in determining the transfer of stress from the matrix to the nanofiller via shear-activated mechanisms. The higher the interfacial shear strength, the greater load it can withstand before interface failure takes place. In the case of poor adherence between the matrix and the nanofillers, the strength of the interface decreases and the particles start acting as defect centers. Hence, strong interaction between the polymer matrix and the nanofiller is mandatory for improving the mechanical properties of the matrix.

It is well known that highly cross-linked polymers, such as epoxies, do not interface well with atomistically smooth carbon nanotube surfaces [41] unless proper functionalization (e.g., attachment of amine groups) of the nanotubes is carried out. For graphene, it is also important to evaluate the quality of the interface and how it may differ from that of carbon nanotubes. Distortions caused by oxygen functionalization and the resultant defects during thermal exfoliation of graphite oxide, as well as the extremely small thickness of GPL, lead to a wrinkled topology at the nanoscale. This nanoscale surface roughness may result in enhanced mechanical interlocking with polymer chains and, consequently, better interfacial adhesion. Such an effect has also been suggested by recent molecular dynamics studies [16, 42].

Raman spectroscopy of the nanocomposite under mechanical loading [43] is a convenient method to study the strength of the interface in nanocomposite materials. Raman spectroscopy is an outcome of inelastic scattering of incident light by a material. Under the influence of a given electric field, the vibrational frequencies of the chemical bonds in a material act as a fingerprint of its interatomic spacing and control the frequency of the emitted photon. Materials that can exhibit a change in their polarization potential show a shift in their Raman peak on varying interatomic bonding distance. In particular, carbon and silicon carbide fibers have shown considerable Raman peak shift on stress application.

Characteristic Raman band shifts have proven to be a powerful technique to study the stress state of carbon-based nanofillers, such as carbon nanotubes and graphene, inside a polymer matrix on application of external stress [44–45]. The stress transfer to the reinforcing filler imposes elastic strains in the filler materials, proportionally shifting their characteristic Raman bands. Tracking the strain-sensitive Raman band-shifts of the nanofiller with respect to the stress state applied to the nanocomposite reveals insight regarding the filler-matrix interactions at the micro-mechanical level. The sensitivity of these strain-induced filler Raman band-shifts can be used to probe the extent of filler-polymer interfacial interactions and determine the basis for their strengthening.

Epoxy polymers are not ideal as a host matrix for Raman study in graphene composites. The reason is that the epoxy structure has a multitude of Raman peaks that lie in close proximity to the characteristic G and 2D band peaks of graphene. This makes it challenging to deconvolute the graphene peaks from the background epoxy peaks. Polymer systems better suited to Raman study are polydimethylsiloxane (PDMS) or polystyrene (PS). In order to disperse graphene in PDMS, the GPL powder is first dispersed in tetrahydrofuran (THF) using a cuphorn sonicator followed by the addition of PDMS. Then, the solvent is removed by heating under continuous stirring on a magnetic plate. Subsequently, a vacuum chamber is used for ~24 hours at ~80°C to remove any leftover solvent. The curing agent is then added in a 1:10 ratio (curing agent to resin). Finally, the mixture is thoroughly blended using a high-speed shear mixer, followed by degassing for ~1 hour and post-cure at ~100°C for ~1 hour. For the PS nanocomposites, both PS and GPL are first dispersed individually in THF using an ultrasonic bath sonicator, followed by mixing them. After homogeneous mixing, THF is filtered out. The leftover solvent is removed using a vacuum chamber for ~48 hours at ~67°C. The dried nanocomposite can be crushed into small particles using a mixer. Using a hot press at ~190°C and ~2 tons of pressure, dog-bone samples can be prepared for the Raman study.

A mini-tensile and compression testing machine operating under a Raman microscope is required for such a study. The standard PDMS peak at 2906 cm^{-1} can serve as an internal calibration, as the PDMS peak does not shift under loading. A mixed Gaussian-Lorentzian profile can be used to fit the Raman peaks of the D (~1350 cm^{-1}), G (~1582 cm^{-1}), D′ (~1620 cm^{-1}), and the G′ (~2700 cm^{-1}) Raman bands. The GPL in the matrix selected for the Raman spectroscopy studies should be > 5 μm below the polymer surface to ensure adequate matrix interaction. Typical results for GPL/PDMS composites are shown in Figure 2.23 [43]. The Raman spectra reveals a shift in the G band of GPL as a function of applied strain; however, after ~7% strain, the peak was found to revert back, close to the original unstrained peak position due to the relaxation of the GPL upon debonding from the matrix.

Figure 2.24 [43] shows the Raman G-band peak shift as a function of the applied strain (both tensile and compressive) for 0.1% weight fraction GPL-PDMS, GPL-PS, and SWNT-PDMS nanocomposites. Figure 2.24(a) shows the results in the elastic regime, while Figure 2.24(b) depicts the response for large strains. In the elastic regime (< 1.5% strain), consistent with the literature, the G-mode peak frequency diminishes (red-shift) with applied tensile strain and increases to higher frequency (blue-shifts) under compressive deformation. The rate of peak shift with strain was ~–2.4 cm^{-1}/composite strain% in ten-

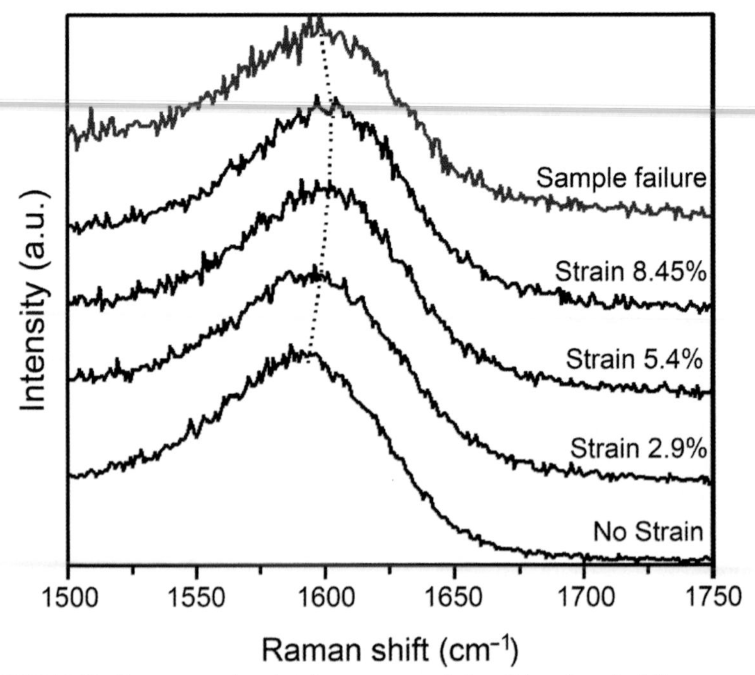

FIGURE 2.23. Raman spectra showing a representative G-band peak shift response of the GPL-PDMS nanocomposite with applied strain. (Adapted with permission from [43]).

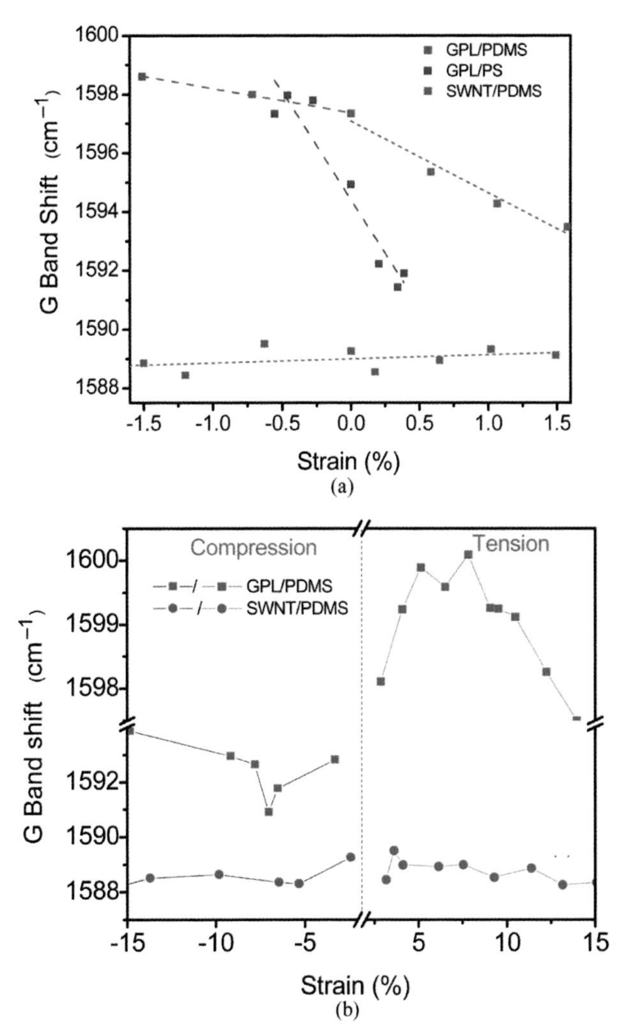

FIGURE 2.24. *(a) A comparison of Raman G-band peak shift of GPL-PDMS, SWNT-PDMS, and GPL-PS nanocomposites up to 1.5% applied strain. Data is shown for both tensile and compressive loading at constant nanonanofiller of 0.1 wt. %. (b) Corresponding Raman G-mode shift for strains > 1.5% for GPL- and SWNT- PDMS nanocomposites. (Adapted from [43] with permission).*

sion and ~1.8 cm^{-1}/ composite strain% under compression. Data is also shown in the figure for 0.1% weight SWNT-PDMS composite; the G-peak shift rate was found to be over an order of magnitude lower (~0.1 cm^{-1}/composite strain %) for SWNT. Since there is a direct correlation between the Raman peak shift and the strain carried by the nanofillers, these results suggest that load transfer at the GPL-PDMS interface appears to be more effective in comparison to the SWNT-PDMS interface. For a higher modulus polystyrene (PS) matrix, the peak shifts were ~7.3 cm^{-1}/composite strain % in both tension and compression for GPL-PS composite in its elastic regime, as shown in Figure 2.24(a). This is consistent with the peak shift rate of monolayer graphene placed on substrates. The lower peak shift in PDMS is attributed to the very low shear modulus of PDMS (~1 MPa) as compared to PS (~1 GPa), a difference of three orders of magnitude. The peak shift rate in PDMS-SWNT composites reported here is much smaller in comparison to epoxy-SWNT composites, which show a shift rate of ~8 cm^{-1}/composite strain % for the 2D peak at ~2640 cm^{-1} for similar reasons.

For large deformations (> 2% strain) in the plastic domain, a surprising observation was that the GPL fillers undergo compression under uniaxial tensile loading and vice versa. Thus, the Raman G-band peak blue-shifts to higher frequencies under tensile load and is red-shifted to lower frequencies under compressive strain, as indicated from a representative sample in Figure 2.24(b). Figure 2.25 shows a best-fit plot for the entire data set, with standard deviations.

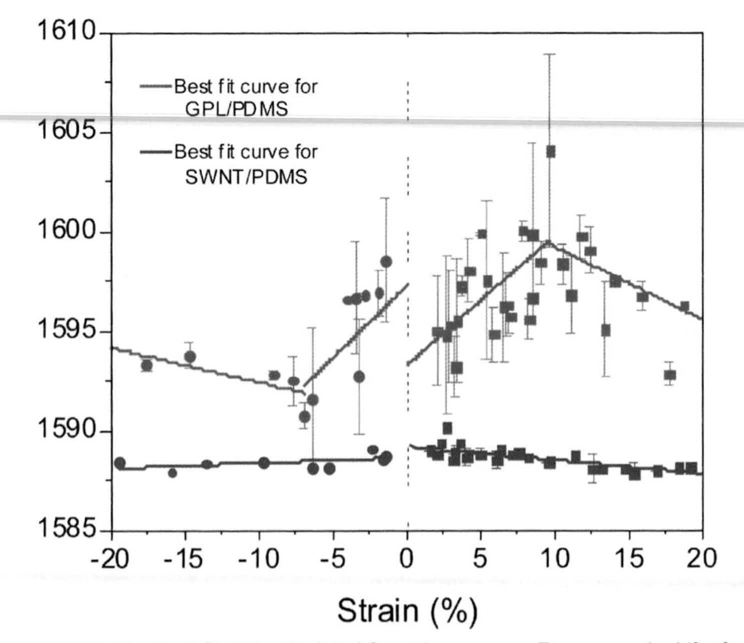

FIGURE 2.25. The best-fit plot calculated from the average Raman peak shifts for the GPL- and SWNT- PDMS nanocomposites, using data set with 10 samples. The data shows the standard deviation of the samples for the best-fit plot. (Adapted from [43] with permission).

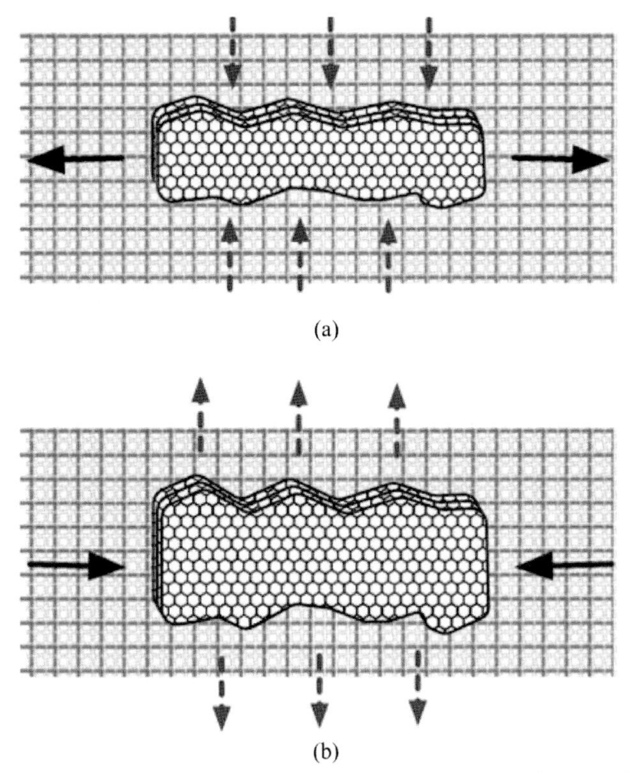

FIGURE 2.26. *Schematic of load transfer mechanism for PDMS-GPL interface under large strains. The drawing of mobile PDMS chains in the direction of applied (a) compressive load, which results in a net tensile strain and (b) tensile load, which results in a net compressive strain on the GPL. (Adapted from [43] with permission).*

The observed deviation of the Raman band shifts in GPL-PDMS nanocomposites at > 1.5% strain probably occurs because, at higher strain, the mobile PDMS chains [with glass transition temperature (–125°C) < room temperature (25°C)] draw in the direction of the uniaxial stress, in contrast to the low tensile strain regime where there is efficient elastic strain transfer from the matrix to GPL [43]. During this process, the chains laterally compress the GPL in its plane, putting the bonds under compression and causing the peak to blue-shift to a higher wave number, as illustrated schematically in Figure 2.26. Similarly, under large compressive strain, the PDMS matrix imparts a net tensile stress to the GPL in the in-plane direction.

Symmetry was observed in the debonding strain for GPL under tensile and compressive stress, which was ~7%, both under compression and tension. Beyond ~7% strain, the G-band peak shift saturates as the

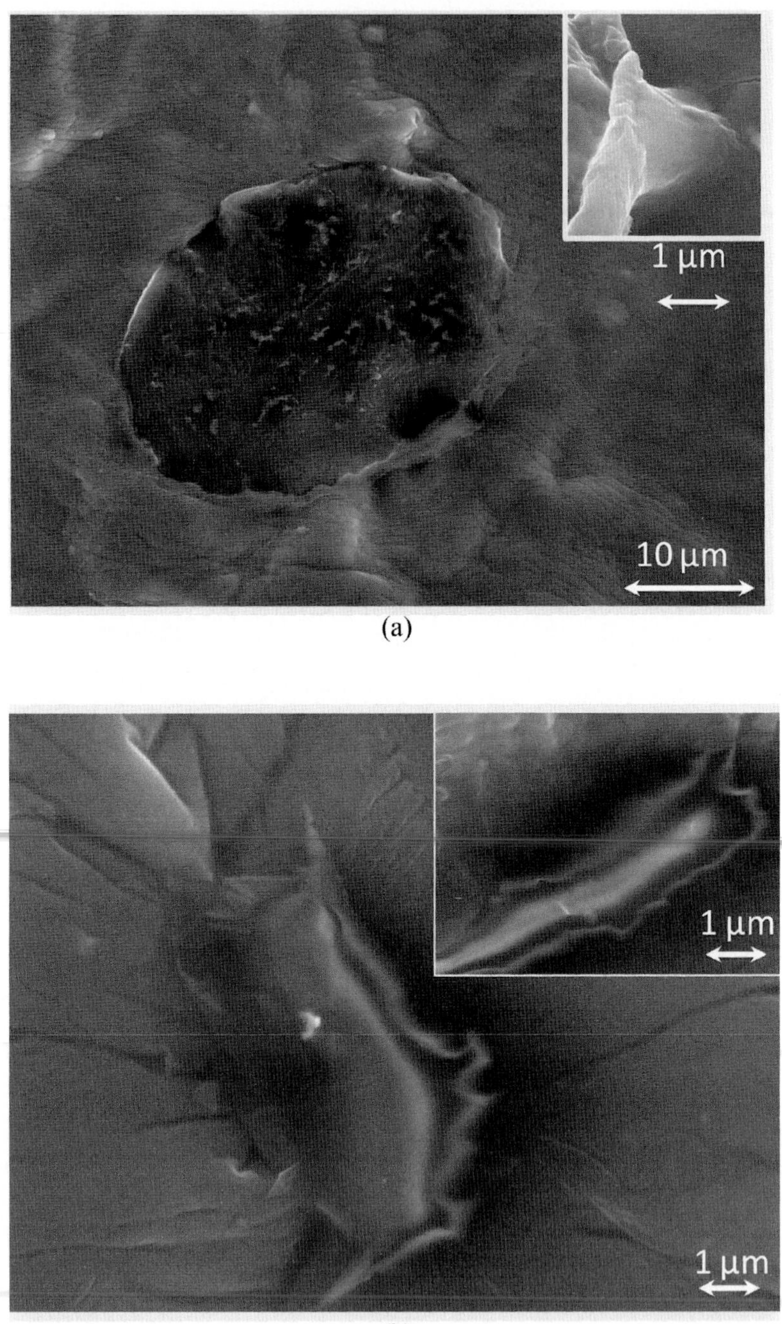

FIGURE 2.27. *SEM of a GPL flake (a) embedded in the unstrained PDMS matrix with the inset showing substantial polymer wetting of GPL, suggesting a strong interface; (b) debonded from PDMS matrix at a strain of ~7%, with the inset showing interfacial debonding of GPL. (Adapted from [43] with permission]).*

GPL interfaces begin to progressively fail and then gradually drop back to the unstrained value. SEM images of the unstrained and strained sample can also be used to study the interface. Figure 2.27(a) shows a SEM image of a GPL embedded in the unstrained PDMS matrix. The inset indicates a high resolution SEM image, revealing the wavy edge character of the graphene wetted by the matrix. As reported above, GPL debonds from PDMS at ~7% strain, as seen in the micrographs of a strained sample; the debonded filler and interface are clearly seen in Figure 2.27(b) and the inset. The observed high strain for debonding is indicative of a strong interface.

Another phenomenon that can be observed is extensive ripple formation (buckling) on baseline (pristine) PDMS and the SWNT/PDMS samples under large compressive strain [43], whereas a smooth surface was observed for GPL/PDMS, as shown in Figure 2.28. This suggests enhanced buckling resistance under compression for GPL as compared to SWNT nanocomposites, which is consistent with the results shown in Section 2.3 of this chapter.

The results summarized in Figure 2.24 indicate that the Raman G-mode peak shift rate in GPL-PDMS composites is at least an order of magnitude higher than the SWNT counterpart at the same loading fraction of ~0.1% by weight. This is verified under both tensile and compressive loading. Since there is a direct correlation between the Raman peak shift and the strain carried by the nanofillers, these results suggest that load transfer at the GPL/polymer interface is far more effective as compared to the SWNT/polymer interface. Therefore, it appears that graphene tends to bind well with a range of polymer matrices, including epoxies, PDMS, and PS. It appears that this effect is related to the

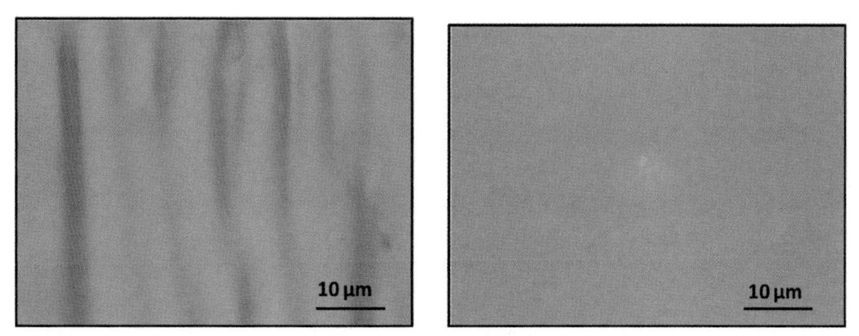

FIGURE 2.28. Optical micrographs of (a) SWNT-PDMS and (b) GPL-PDMS sample surface under large compressive strain (~15%). For the SWNT nanocomposite a rippled surface texture is seen, indicating extensive local buckling in contrast to the GPL nanocomposite, where no such effect is observed. (Adapted from [43] with permission).

rough and wrinkled surface texture of graphene, which leads to strong mechanical interlocking with the polymer matrix.

2.8. CHARACTERIZING THE INTERPHASE IN GRAPHENE POLYMER COMPOSITES

An important feature of nanocomposite materials is that the nanofiller surface affects the dynamic properties of the polymer matrix in its close proximity, resulting in the formation of an "interphase" region [46–49]. There have been reports of significantly altered nanocomposite properties due to interphase formation near nanofillers, as they have large specific surface area to affect the properties of the host matrix. The presence of fillers can affect the crystallinity, mobility, and other structural aspects of the polymer chains. However, it is important to note that contradictory results have been found in different polymer systems. For instance, nanofillers have been found to restrict the polymer chains in some composites while they enhance the mobility in others. Differential scanning calorimetry to determine the glass transition temperature, as well as atomic force microscopy to map the stiffness of the material in the interphase region, are some of the commonly used techniques for studying the interphase in nanocomposites.

For differential scanning calorimetry (DSC), a sample weight of ~5 mg is sealed in a hermetic aluminum pan. The temperature run is carried out at a ramp rate of ~10°C.min^{-1}, from 0°C to ~250°C. To obtain the second heat flow plot, the sample is cooled and the experiment is repeated. The glass transition temperature (T_g) can be obtained from the DSC plots of the second heat flow run, as shown in Figure 2.29. An increase in T_g of ~8°C is obtained for ~0.2 wt.% of GPL in epoxy, as shown in Figure 2.30(a). At a higher weight fraction of GPL (> 0.2 wt.%), the T_g diminished due to agglomeration. The relaxation peak width for the neat polymer and nanocomposites can be calculated from the DSC plots, given by $\Delta T = T_U - T_L$, where T_U and T_L are the upper and lower bounds of the tangents as shown in the inset to Figure 2.29. As seen in Figure 2.30(b), the relaxation peak width increased significantly with increasing graphene content in the epoxy matrix.

The increase in the T_g for the nanocomposite is attributed to the improved filler-matrix adhesion and the likely conformal changes that epoxy chains undergo in the vicinity of graphene. The reduced mobility of the polymer chains in the filler vicinity expectedly raises the T_g of the composite. The result is an indication of good wetting and interfacial

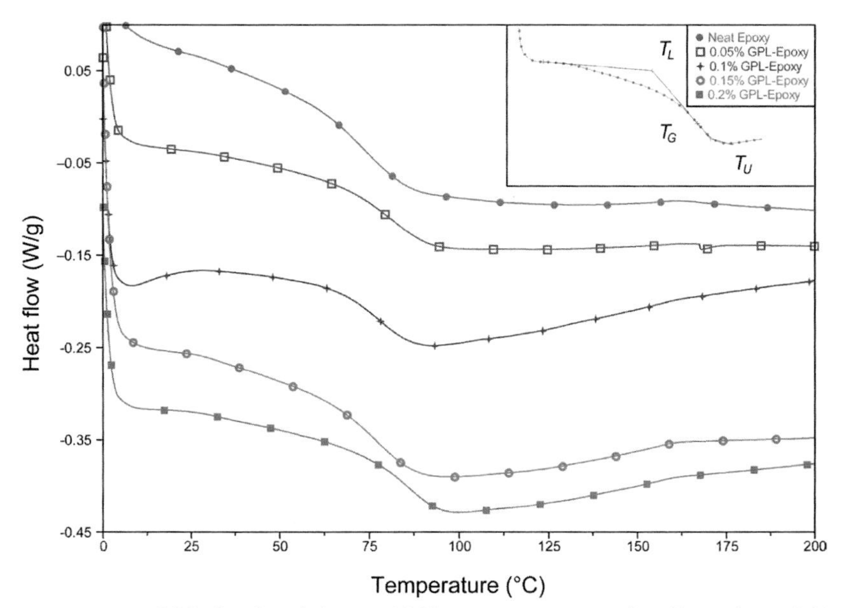

FIGURE 2.29. *DSC plots for pristine- and GPL-epoxy nanocomposite with varying weight fraction, with the inset showing an illustration for calculation.*

adhesion of wrinkled graphene sheets by the polymer and is consistent with the enhancement in mechanical properties observed for graphene/ epoxy nanocomposites. The broadening of the relaxation peak is an indication of heterogeneity in the material, as different clusters have different responses. The fillers that have the tendency to immobilize the vicinity materials lead to non-cooperative motion of the polymer chains

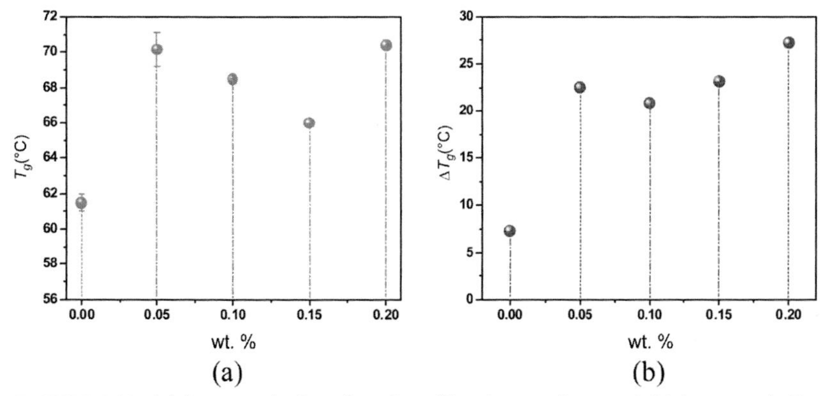

FIGURE 2.30. *(a) Increase in the glass transition temperature and (b) increase in the relaxation peak width with increasing wt. % of GPL in epoxy.*

and broaden the range of temperature in which transition takes place. The phenomenon is attributed to the tendency of GPL to affect the polymer chains in its vicinity, an effect that increases with filler content. The increasing relaxation peak width in the nanocomposite is indicative of the increasing temperature range required to increase the mobility of the polymer chains. Thus, it appears that the graphene fillers diminish the polymer chain mobility in their vicinity and, hence, increase the temperature range required by them to gain complete mobility. To compare the increase in the relaxation peak width for plain multi-walled carbon nanotubes (MWNTs), amine-treated MWNT (A-MWNT), and GPL-epoxy nanocomposites, 0.2 wt. % of filler loading was chosen and the DSC scans were repeated. The percentage change in the relaxation peak width is shown in the bar chart of Figure 2.31. As can be seen from Figure 2.31, the ΔT peak width of the GPL nanocomposite is ~3.5 times that of neat epoxy, as well as MWNT nanocomposite, and ~3 times that of A-MWNT nanocomposite. From these findings, it can be inferred that the mobility of polymer chains is more restricted around GPL as compared to other forms of nanofillers. This could be due to their wrinkled and rough surface texture that facilitates mechanical interlocking with polymer chains, as well as due to the abundance of structural and topological defects in graphene that provide chemical handles for covalent interaction with epoxy groups.

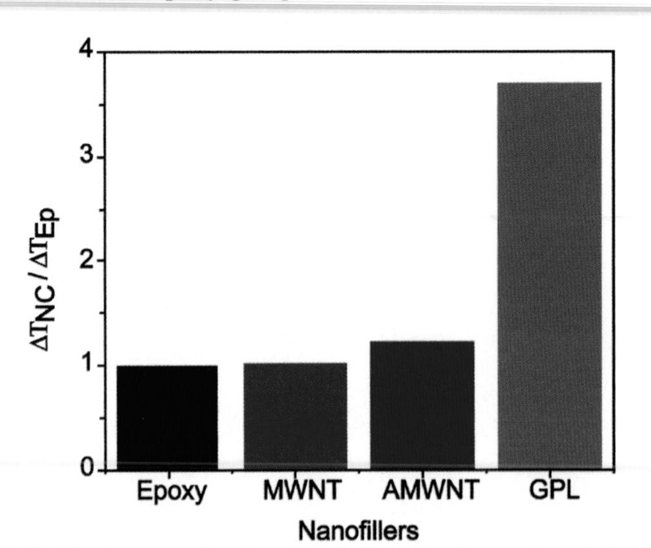

FIGURE 2.31. *Showing normalized increase in the relaxation peak width for the nanocomposite with 0.2 wt. % of neat epoxy, MWNT-, A-MWNT- and GPL- epoxy nanocomposite obtained by DSC.*

Atomic force microscopy (AFM) is a powerful nano-characterization tool to study the topography and mechanical properties of composite materials. AFM has proven useful in characterizing the properties of the interphase formed between a filler and the matrix. Atomic force micrographs can be used in various modes with varied data processing techniques to extract a wealth of information; e.g., tapping mode AFM can give topographical information (height image) and quantitative material stiffness (phase images). A phase contrast between the bulk matrix, the filler, and the interfacial matrix is a qualitative way to probe the localized influence of fillers on the matrix. The principal factor affecting phase contrast is the variation in the interactions between the filler and the probe-tip.

AFM study typically requires a very flat surface to ensure that excessive height perturbations do not affect the accuracy of the measurements. Consequently, microtomed samples of GPL-polymer nanocomposites must first be obtained using glass knives. The microtomed samples must then be polished using standard metallographic techniques using different grades of polishing papers, followed by a final polish with 0.1 μm size diamond paste to obtain samples with R_a (surface roughness) < 1 μm. Next, focused ion beam (FIB) milling using gallium ions can be used to obtain sections of the samples with roughness < 10 nm. The atomic force microscopy (AFM) measurements can then be performed on such ultra-smooth specimens using a scanning probe microscope. Cantilever-beam silicon tips and/or carbon nanotube AFM tips with tip diameter of 5–8 nm are typically used for such experiments. Figure 2.32 is an AFM micrograph of a GPL-epoxy nanocomposite sample showing the GPL embedded in the epoxy matrix. The bright region (with high phase angles) corresponds to GPL and the darker region (with low phase angles) corresponds to the epoxy matrix. The corresponding height image is shown in the inset in Figure 2.32. To investigate the formation of an interphase, multiple AFM scans can be performed on various regions of the sample and high magnification scans can be done in areas displaying significant phase differences. A representative scan of such a region is shown in Figure 2.33. The transition from the high stiffness graphene regions to the low stiffness epoxy region is clearly not abrupt and, in many regions, a gradation in stiffness is discernable. This may be a result of immobilization of polymer chains to the graphene surface, resulting in enhanced stiffness of the interfacial polymer region. Figure 2.34 shows high resolution AFM scans of this interphase region using an AFM probe instrumented with a multi-walled carbon nanotube tip.

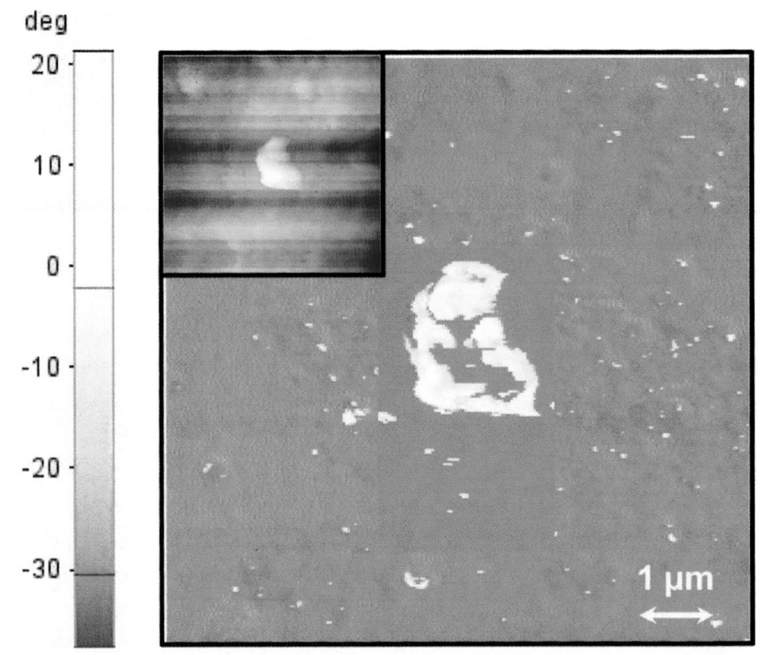

FIGURE 2.32. Phase image of a graphene platelet embedded in epoxy matrix. The inset is a corresponding height image of the same region.

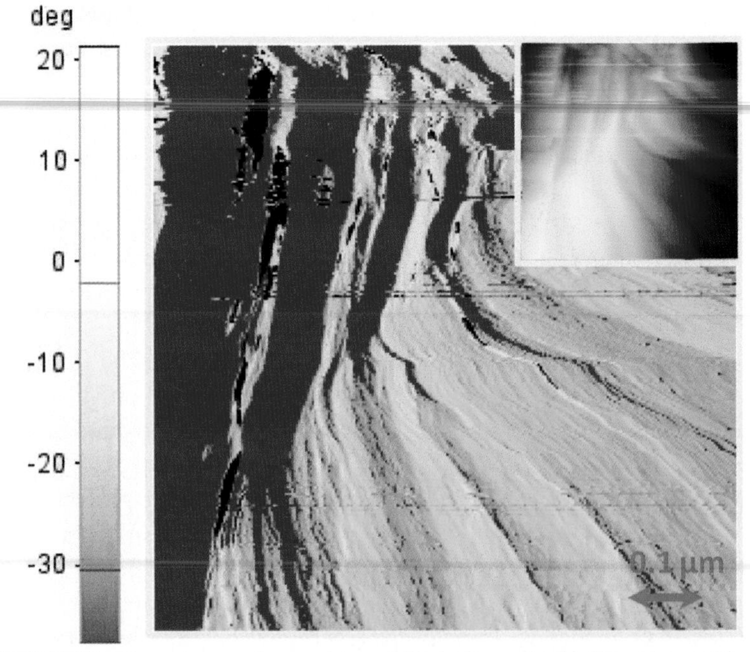

FIGURE 2.33. Phase image of a graphene platelet edge embedded in epoxy matrix. The bright region corresponds to GPL and the dark region to matrix. The inset is the corresponding height image.

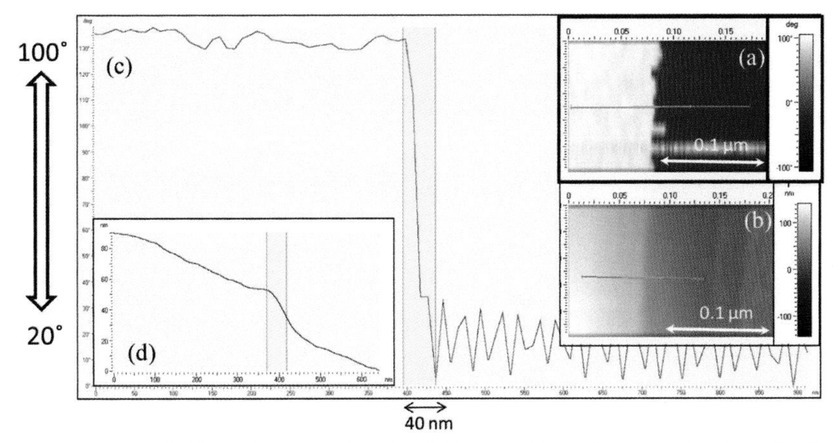

FIGURE 2.34. (a) Phase image and (b) height image of the GPL-epoxy interface. While the height profile along GPL varies from ~40–80 nm, the phase angle maintains a constant value of ~100º. (c) Phase profile showing the sharp 80º drop in the phase angle across the GPL towards the epoxy matrix over a distance of ~40 nm. (d) Height profile indicating the 60 nm gradient on the GPL. The green lines indicate the sample length along which the profile was measured.

The nanotube tips enable a spatial resolution of about 10–20 nm. A steep decrease in the phase angle of ~80º (from ~100º to ~20º) within a distance of ~30–40 nm can be observed as the AFM tip is scanned across the GPL-matrix interface. Given that the resolution of the AFM probe is about 10 nm, it can be concluded that an interphase region of the order of a few tens of nanometers likely exists, in which the local stiffness of the epoxy is significantly higher than the bulk epoxy stiffness. The formation of the interphase may be related to chemical bonding between epoxy groups and defect sites that are ubiquitous on the graphene sheet surface and the sheet edges. Such interactions will severely restrict epoxy chain mobility near the graphene inclusions and this also explains why the glass transition temperature for the epoxy is significantly elevated due to the graphene additives.

2.9. VISCOELASTIC PROPERTIES

Another important property of polymers is their viscoelastic nature [50]. For dynamic stress application at high frequencies, a lag between stress and strain is created given by a phase angle δ, which lies between, $0 < \delta < 90^\circ$. The tangent of δ is defined as the ratio of storage to loss modulus. The storage modulus, $E'(\omega)$, and loss modulus, $E''(\omega)$, characterize the viscoelastic behavior of the material system at the frequency of stress application (ω). For example, in case of a sinusoidal stress of the type:

$$\sigma(t) = \sigma_0 \sin(\omega t) \tag{51}$$

The strain response of a viscoelastic material is given by:

$$\gamma(t) = E'(\omega)\sin(\omega \times t) + E''(\omega)\cos(\omega \times t) \tag{52}$$

where, E' is obtained from the ratio of in-phase stress to strain and E'' is obtained from the ratio of out-phase stress to strain. Tan δ is given by:

$$\tan\delta = E''/E' \tag{53}$$

$$\delta = \tan^{-1} E''/E' \tag{54}$$

where tan δ is the measure of the viscoelastic damping in the material, which quantifies the internal energy losses in the material due to friction.

For viscoelastic characterization, polished dog-bone shaped samples (typical size ~8 mm × ~3 mm × ~0.5 mm) are used. The samples can be tested in the tensile mode (e.g., using a Rheometer DMTA V). The measurements include strain sweeps to study the storage modulus, loss modulus, and loss tangent for pristine epoxy- and graphene- epoxy nanocomposites (~0.125 wt.% of GPL) [51]. The damping and thermo-mechanical properties are generated by performing strain sweeps from 0 to 0.5% at room temperature (24°C and 10 Hz). A representative plot of storage, loss modulus, and loss tangent for neat epoxy as a function of strain amplitude is shown in Figure 2.35(a). The strain amplitude does not have a significant impact on the viscoelastic properties of the material in the 0–0.3% strain amplitude range. On comparing the plots for neat epoxy and graphene-epoxy nanocomposite [Figure 2.35(b)–(d)], loss tangent is seen to decrease by a factor of ~1.5 with increasing strain amplitude. An ~1.5 fold increase in the storage modulus, from ~2.2 GPa to ~3.4 GPa, was also observed, which was maintained throughout the strain sweep from 0.005% to 0.3%. The reduced loss tangent is a direct outcome of increased storage modulus of the nanocomposite. The enhanced storage modulus implies strong load transfer at the graphene-epoxy interface. For the loss modulus, no significant change was observed. This again indicates a strong interface. If the interface were weak, one would expect interfacial slippage, which would result in frictional energy dissipation, which would enhance the loss modulus and the loss factor. Such behavior is observed in carbon nanotube composites for polymers that do not interact strongly with the nanotubes. In the case of graphene, no such effect is seen, which corroborates the Raman measurements, indicating a strong interface.

Figure 2.36(a) shows a representative plot of the storage modulus, loss modulus, and loss tangent for a neat polydimethylsiloxane (PDMS) sample as a function of strain amplitude. Figure 2.36(b)–(d) shows the comparative results for loss tangent, storage modulus, and loss modulus for neat PDMS and graphene-PDMS (0.125 wt. %) nanocomposite with strain amplitude sweep.

Similar to the epoxy case, the loss tangent of the neat PDMS is larger by a factor of ~1.5 compared to the GPL-PDMS nanocomposite and this difference was maintained up to ~5% strain amplitude. A 1.5

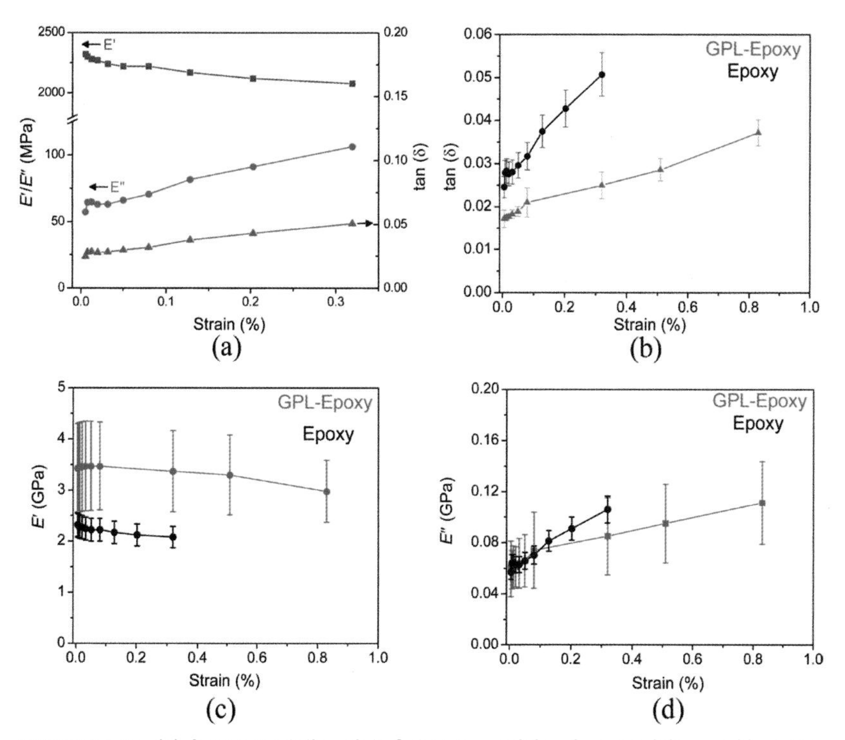

FIGURE 2.35. *(a) A representative plot of storage modulus, loss modulus, and loss tangent for a neat PDMS sample with sweeping strain amplitude. (b) Loss tangent, (c) storage modulus, and (d) loss modulus for neat PDMS and 0.125 wt. % GPL-PDMS nanocomposite. (Adapted from [51] with permission).*

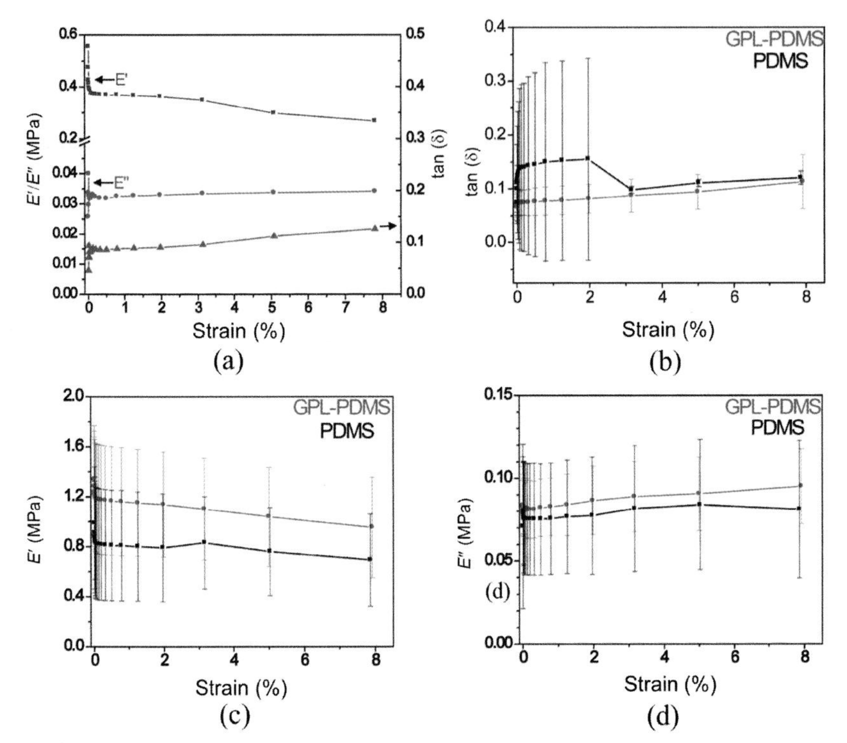

FIGURE 2.36. (a) Representative plot of storage modulus, loss modulus, and loss tangent for a neat epoxy sample with sweeping strain amplitude. (b) Loss tangent, (c) storage modulus, and (d) loss modulus for neat epoxy and 0.125 wt. % GPL-epoxy nanocomposite. (Adapted from [51] with permission).

times increase in the storage modulus at 0.008% strain amplitude in the nanocomposite over neat PDMS was observed, which again confirms good load transfer at the graphene/PDMS interface. The loss modulus of the GPL/PDMS nanocomposite was marginally greater than the neat PDMS. The decrease in the loss tangent for the GPL/PDMS composite compared to the neat PDMS is a result of the enhanced storage modulus of the graphene/PDMS composite compared to the pristine PDMS. At ~5–7% strain amplitude, the loss tangent of the GPL/PDMS and the baseline PDMS samples is nearly the same, which is expected given that graphene debonds from the PDMS in this strain range (see Raman results in Figure 2.25).

2.10. WEAR PROPERTIES

Development of low friction, wear-resistant surfaces is a classical problem in tribology and surface engineering. While fluid film (hydrodynamic lubrication) provides the lowest friction coefficients, there are many practical situations where fluid lubrication of a sliding system may not be feasible [52]. For example, in space applications, extremely low temperatures may cause conventional liquid lubricants to freeze, and high temperatures may cause them to degrade or evaporate, which may also occur in vacuum environments. Other applications may not be able to tolerate contamination by a fluid lubricant, or it may simply be desired to avoid the complexity, weight, and expense of the recirculation system for fluid lubrication [52].

When fluid lubrication is not practical, solid lubricants are often used. One of the most common solid lubricants being utilized by industry is Polytetrafluoroethylene (PTFE), popularly called "Teflon" (its DuPont brand name). PTFE is a thermoplastic material with wide ranging applications in the aerospace, chemical-processing, medical, automotive, and electronic industries [52]. The factors that afford PTFE such widespread applicability include its exceptional ability to provide low friction, high chemical inertness; its thermal stability over a wide range of temperatures; its high melting point (327°C); and its good insulating and dielectric properties. Figure 2.37(a) shows the side and end views of the helical conformation that the PTFE molecule adopts under ambient low pressure conditions.

While PTFE provides low friction coefficients during dry sliding, its exceedingly high rate of wear limits its performance and hinders its applications. For example, at a temperature of ~23°C and sliding speeds > 8 mm/s, PTFE wears at a rate as high as 10^{-3} mm^3/Nm, with plate-like debris that is several micrometers in thickness and hundreds of micrometers in in-plane dimensions [53]. It is well known that incorporating hard micron-scale fillers in the PTFE matrix makes the composite wear-resistant when compared to unfilled PTFE. The hypotheses for the role of micron-scale fillers in the wear mechanism of PTFE composites are many and varied, and sometimes conflicting in nature. The mechanisms that appear to capture most of the experimental observations were ones in which the microfillers reduced wear by regulating the debris size [54–55]. These mechanisms suggest that nano-scale fillers, unlike micro-scale fillers, would perhaps lack the size scale necessary to interfere with the large, lumpy debris generation processes in PTFE and, therefore, would be ineffective in reducing wear. However, as seen in Figure 2.37, in the last decade nanofillers have been found to not only be capable of providing wear resistance comparable to microfillers, but in some instances even better, by up to two orders of magnitude. Nanofillers have been thought to reduce wear by a number of mechanisms, including increasing fracture toughness and preventing large scale destruction in PTFE; enhancing transfer film-countersurface adhesion; or causing structural phase changes to the PTFE matrix, making it more wear resistant.

Graphene is intriguing from the point of view of wear suppression since it is a multi-scale filler with micro-scale sheet dimensions of or-

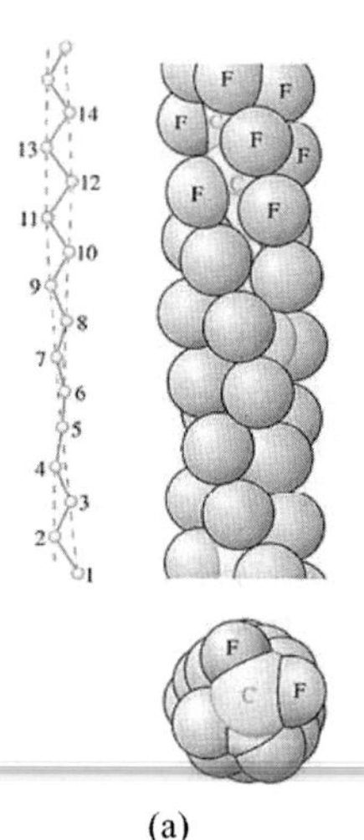

(a)

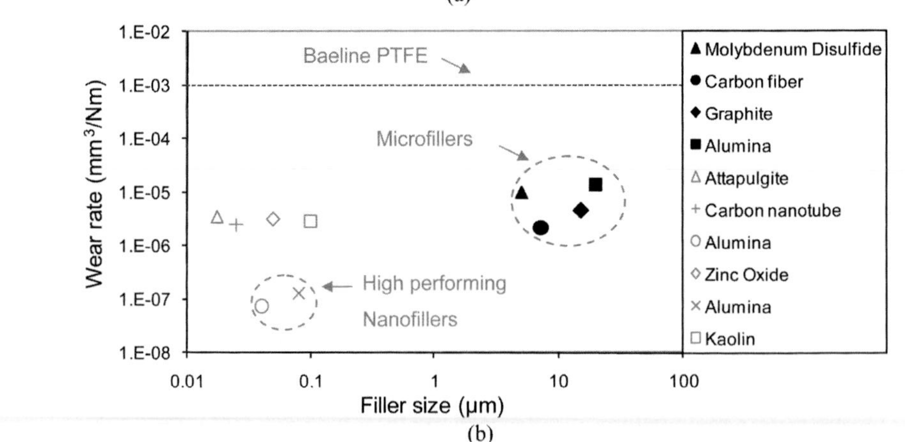

(b)

FIGURE 2.37. *(a) Chemical structure of Polytetrafluoroethylene (PTFE) or Teflon. (b) Wear rates from literature of PTFE incorporating the following micro- and nanofillers: molybdenum disulfide, 7.2 μm diameter carbon fiber, graphite, 20 μm alumina, acid-treated attapulgite, 20 to 30 nm diameter multi-walled carbon nanotube, 40 nm alumina, Zinc Oxide, 80 nm alumina, and kaolin. (Adapted from [56] with permission).*

der of ~5 μm × 5 μm coupled with nanometer scale sheet thickness. The micro-scale dimensions of the graphene sheets might enable it to effectively interfere with the debris generation processes in polymers, while the nanometer scale thickness, low density, and the planer sheet geometry of graphene generate a huge interfacial contact area with a very large number density of graphene sheets in the polymer matrix. Moreover, sliding of individual graphene planes within few-layered GPLs might further enhance the lubrication effect. In this section, some preliminary results are described that indicate graphene shows extraordinary potential as a wear-suppressant additive for PTFE. Incorporation of ~10% weight fraction of graphene in PTFE reduces the wear rate by 4 orders of magnitude from ~10^{-3} mm^3/Nm to ~10^{-7} mm^3/Nm.

PTFE granular resin, with an average particle size of 30 μm (DuPont 7C), can be utilized as the matrix material. The first step in polymer composite preparation consists of dispersing the filler particles in the matrix powder. Hauschild mixing is the primary mixing technique that can be used for this purpose. Five mixing cycles, each lasting 30 seconds, are required for each powder mixture at a main platform speed of 1300 RPM. About 5–6 g of the mixed powders are then transferred to a cylindrical die and pressed at 40 MPa for ~15 minutes at room temperature to form cylindrical pucks. The pucks are then sintered by heating at a rate of ~100°C/hr to ~360°C, where they are held for ~3 hours. The pucks are then cooled to room temperature at approximately ~100°C/hr. To maintain an inert oxygen-free atmosphere during the sintering process, nitrogen gas is flowed through the chamber at a rate of 4 l/min. Figure 2.38 shows a typical scanning electron microscopy (SEM) image of the freeze-fractured surface of the PTFE/graphene composite for ~2% weight fraction of GPL additives. There is no indication of large agglomeration of GPL. High resolution SEM (inset in Figure 2.38) shows an individual GPL filler embedded in the PTFE matrix; the wrinkled surface texture of the GPL is clearly discernable in the inset image. Similar results were obtained at GPL weight fractions of up to 10%. These results indicate that a reasonable dispersion of the GPL in the PTFE matrix is possible.

The sintered PTFE composite pucks can be machined by milling into pins with a square cross-section of sides ~4 mm × ~4 mm and a typical length of ~10 mm, with the square cross-section to serve as the contact surface during wear testing. Stainless steel 304 with average surface roughness of 0.015–0.03 μm is typically used as the counter-surface material. The wear tests can be conveniently performed on a multi-sta-

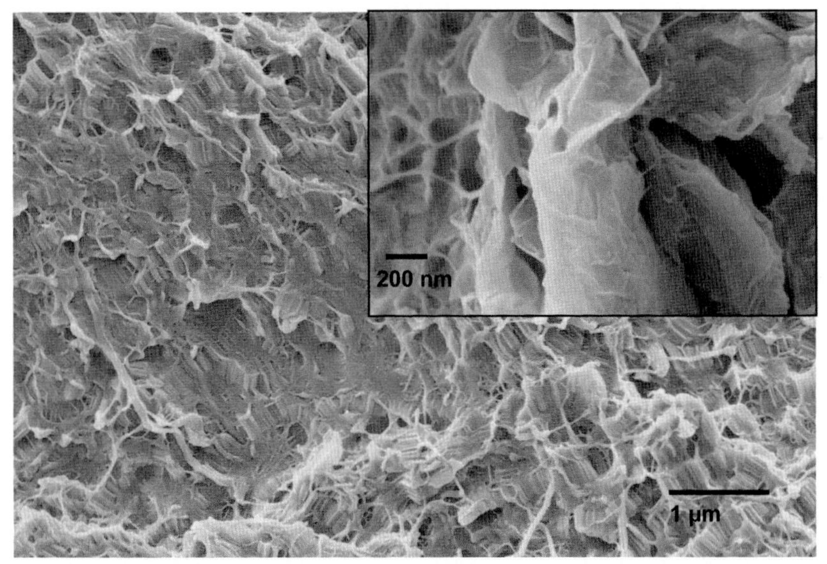

FIGURE 2.38. SEM image of freeze-fractured 2% weight GPL/PTFE composite. Inset shows a typical GPL in the matrix. (Adapted from [52] with permission).

tion tribometer in a reciprocating pin-on-flat configuration in ambient air at room temperature. Pneumatic loading can be applied to the pin holder to generate a contact pressure (~3.125 MPa) at the pin-countersurface interface. Such tests are typically performed at sliding speeds in the range of ~10 cm/s. The sliding must be stopped periodically and mass loss-based wear measurements of the composite pins need to be performed. The mass losses can then be converted to wear volumes using the composite densities. The value of wear volume per unit sliding distance for each composite is then obtained by performing a linear regression analysis of the steady-state linear portion of its wear volume versus sliding distance record, which is obtained following the initial transient run-in period. This value, divided by the normal load, yields the steady-state composite wear rate with units of mm^3/N-m.

Figure 2.39 shows results for wear volume plotted against the sliding distance for the unfilled PTFE and for various weight fraction of GPL additives. A dramatic decrease in wear volume is evident as the GPL weight fraction is increased from 0.32% to 5%. Clearly, the GPL additives have significantly enhanced the wear resistance of the neat PTFE matrix. Results for the steady state composite wear rate for the baseline PTFE and the PTFE/GPL nanocomposites are summarized in Figure 2.40. The baseline PTFE displays a wear rate of $\sim 0.4 \times 10^{-3}$ mm^3/Nm,

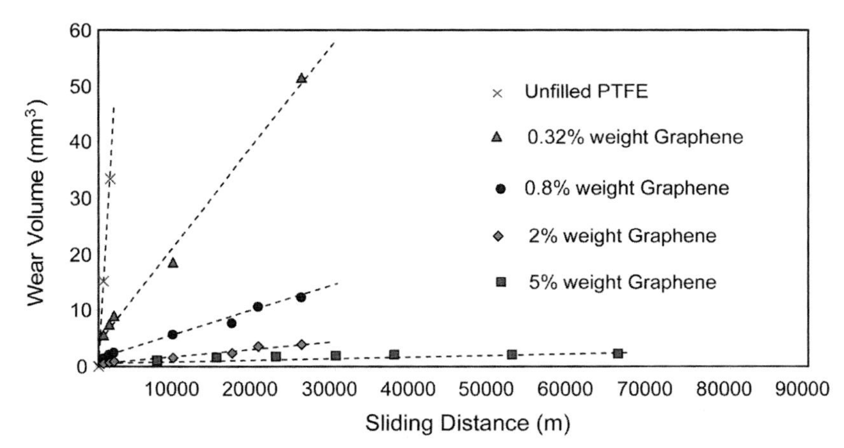

FIGURE 2.39. Wear volume vs. sliding distance for baseline PTFE and PTFE/Graphene composites. (Adapted from [52] with permission).

which is consistent with the literature. Until about 0.1% weight of GPL, there is no significant impact on wear rate. After this threshold level, the wear rate decreases continuously by four orders of magnitude to $\sim 1 \times 10^{-7}$ mm^3/Nm as the GPL weight fraction is increased to 10%. Importantly, even at 10% there is no indication of saturation, which suggests that the wear rate could potentially be decreased further by increasing the GPL weight fraction above 10%. The wear rate suppression for PTFE achieved using GPL is comparable with the results

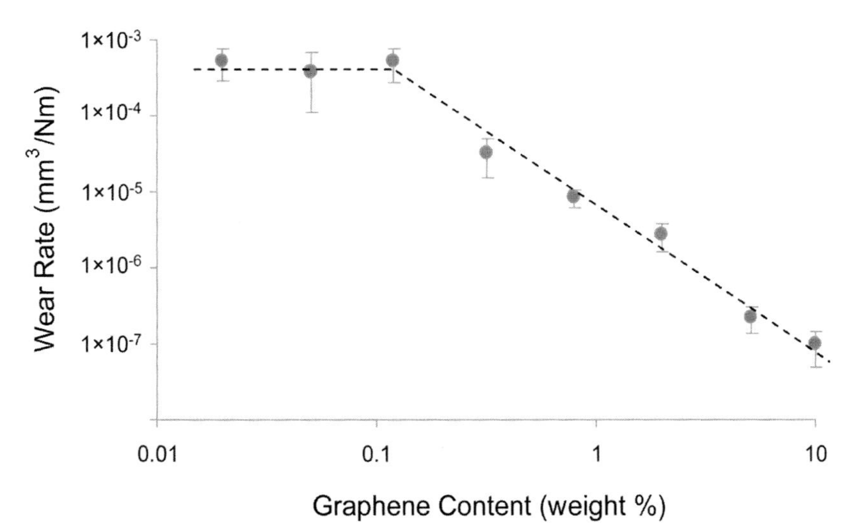

FIGURE 2.40. Steady state wear rate for PTFE/graphene composites. (Adapted from [52] with permission).

achieved for the best available nanofillers (see Figure 2.37). Further, based on the trends seen in Figure 2.40, it appears that GPL can outperform all other categories of nanofillers and microfillers provided the GPL weight fractions are increased beyond 10%. These results are very encouraging and demonstrate the dramatic impact of graphene additives on wear suppression in PTFE composites. At the present time there is no in-depth understanding on the underlying mechanisms responsible for wear suppression in graphene composites. However, the evidence so far suggests that crack deflection may again be the dominant mechanism. Wear originates as a result of propagation at a depth of a few micrometers of sub-surface cracks parallel to the sliding surface, with debris formation occurring when these cracks eventually connect with the surface. The presence of high aspect ratio graphene sheets will cause such sub-surface matrix cracks to be deflected through a tortuous path, thereby boosting the wear resistance of the polymer.

2.11. CREEP

Creep is the time-dependent continued deformation of a structure under a persistent load that is usually significantly lower than the material's yield stress. Materials with high creep resistance are critical in long-term structural applications to ensure dimensional stability and load carrying capability. Creep in thermosetting epoxy polymers is important since they are used in a variety of applications, such as coatings, adhesives, industrial tooling, and composites, as well as in the semi-conductor and electronics packaging industries. Epoxies consist of long molecules that are cross-linked and contain chains of atoms held together by covalent bonds. The chemical bonding is strong and directional along the chains, but adjacent chains are bonded by weak Van der Waals forces. Under persistent external loading, the molecular chains are stretched and re-oriented to carry the load, resulting in mechanical creep. The creep deformation is strongly dependent on the stress level and the temperature, as well as the microstructure of the material.

In this section, the effect of graphene platelets (GPLs) on creep in epoxies is described. Classical macroscopic creep, as well as the characterization of creep by nano-indentation, are discussed [57]. The macro-scale creep tests are performed using an MTS-858 servo-hydraulic test system operating in constant force mode. The force is measured by the load cell of the MTS system and an internal feedback circuit is used to maintain constant stress in the sample. The time dependent strain response of the sample is measured using an extensometer (MTS 632.26E-20) attached to the sample. Figure 2.41 shows the creep response of the baseline and nanocomposite epoxies with ~0.1%, ~0.3%, and ~0.5% weight fraction of GPL additives. The tests are performed at constant stress levels of ~20 MPa and ~40 MPa at room temperature. These stress levels are smaller than the yield strength of the samples. At 20 MPa stress, the creep deformation is very small and there is no great difference between the various samples. However, at 40 MPa stress, the ~0.1 wt. % GPL/epoxy nanocomposite shows a significant reduction in creep strain compared to the neat epoxy. By contrast, creep in the ~0.3% and ~0.5% GPL/epoxy samples is greater than in the baseline epoxy. Poor dispersion and agglomeration of GPL fillers is believed to be the reason for the degradation in creep response at higher weight fractions, which is consistent with the prior observations made in Section 2.4 regarding poor dispersion of GPL above a weight fraction of ~0.125%.

In addition to macro-scale testing, the nano-indentation technique

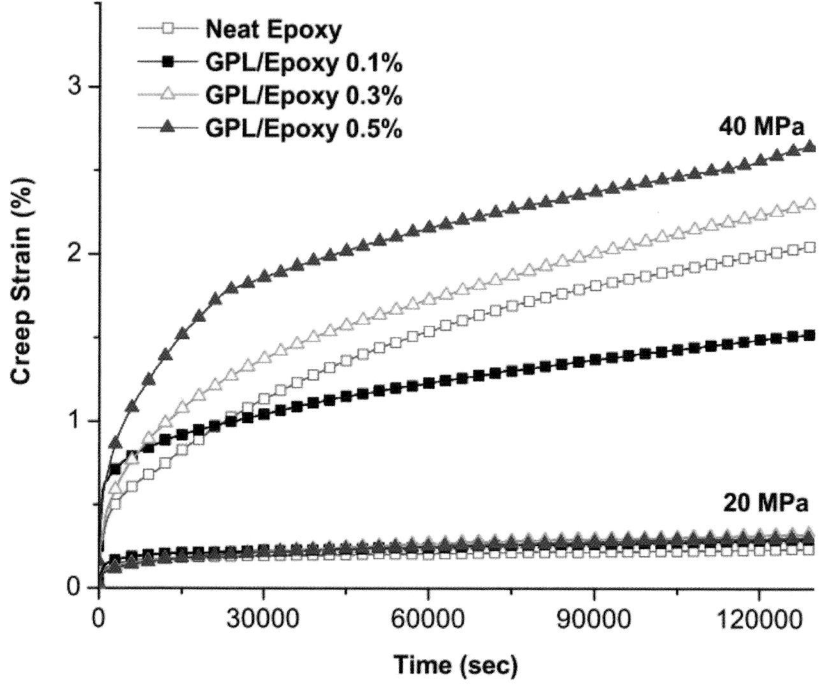

FIGURE 2.41. *Macroscopic creep testing. Strain vs. time plots obtained from creep tests conducted at 20 MPa and 40 MPa tensile stress at room temperature. (Adapted from [57] with permission).*

can also be utilized to characterize the material's creep behavior. For this purpose, a nanocomposite sample that has a weight fraction at which the most improvement in macroscopic creep behavior (~0.1 wt. % GPL-epoxy nanocomposite) was chosen and its creep properties were compared with a pristine epoxy sample. Each test includes loading and unloading sections with identical loading/unloading rates and a ~600-second hold period at the maximum load. To characterize creep response, the creep displacement, d^c, is defined as the difference of the indentation depth during the hold period, $d(t_c)$, and the indentation depth at the beginning of the hold period, $d(t_0)$.

$$d^c = d(t^c) - d(t_0^c); \quad t_0^c = 0 \text{ sec} \tag{55}$$

Creep time, t^c, is defined as the difference of time during the hold period, t, and the time at the start of the hold period, t_0 (e.g., the creep time at the end of the hold period is 600 seconds.)

$$t^c = t - t_o \tag{56}$$

Tests are performed in load control mode and conducted using a maximum load of 2 mN. The load vs. indentation depth plots, as well as associated creep displacement vs. creep time plots, are shown in Figure 2.42. Data is shown for indentation performed at different positions on the sample. The results indicate a significant decrease in the creep displacement rate for the 0.1 wt. % nanocomposite compared to the pristine epoxy. In addition, one of the creep curves in Figure 2.42(b) shows an order of magnitude lower creep rate (~1.175×10^{-4} nm/sec) than the baseline epoxy (creep rate ~2.098×10^{-3} nm/sec). This outlier in the local creep response is evidence of significant heterogeneity in the nanocomposite samples, leading to regions where the local creep response is greatly suppressed.

It is interesting to investigate the effect of temperature on the creep response, while keeping the stress level low (20 MPa). To this end we consider the 0.1 wt. % GPL system. Figure 2.43(a) shows macroscopic test data for the creep strain versus time for both pristine and nanocomposite samples at room temperature and at T = 23°C, 40°C, and 55°C. At room temperature, the curves corresponding to the two materials overlap since the stress levels are intentionally kept very low (~20 MPa). However, differences are seen as the temperature increases in spite of the relatively low stress levels. For example, at 55°C the epoxy creeps significantly more than the nanocomposite. Figure 2.43(b) compares the creep performance of GPL with SWNT and MWNT fillers at a constant nanofiller weight fraction of ~0.1% and a temperature of ~55°C and ~20 MPa stress. The results indicate that both epoxy-carbon nanotube systems creep at the same rate with neat epoxy at 55°C. The

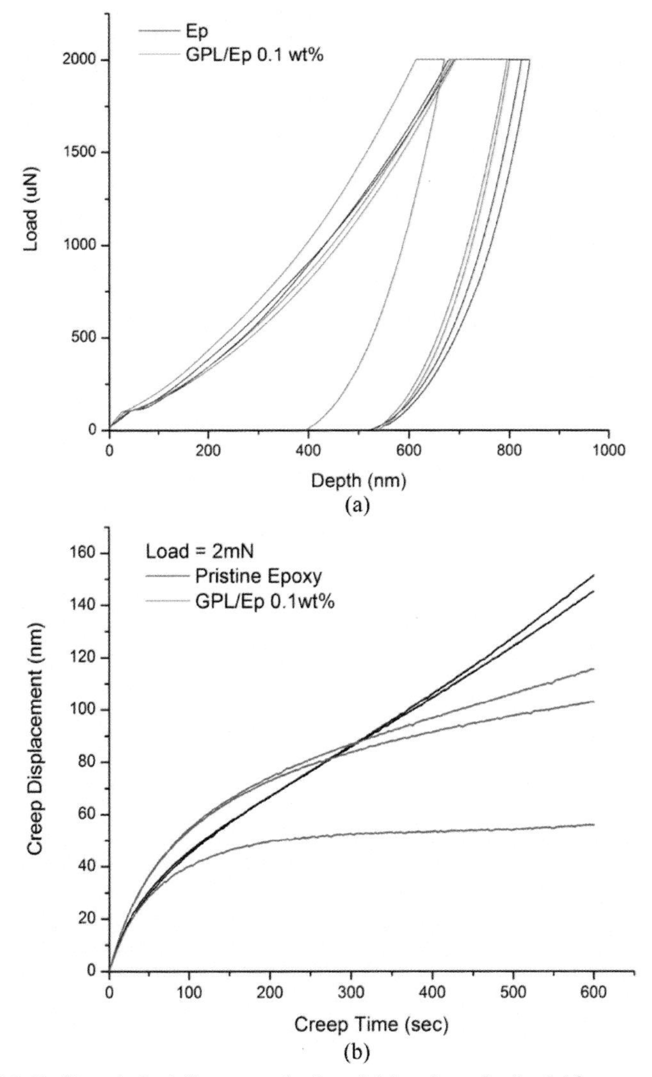

FIGURE 2.42. *Nano-indentation creep testing. (a) Load vs. depth plot for nano-indentation tests on pristine epoxy and 0.1 wt. % GPL/epoxy samples with maximum load of 2 mN. (b) Creep displacement vs. creep time during the hold period at the maximum load. (Adapted from [57] with permission).*

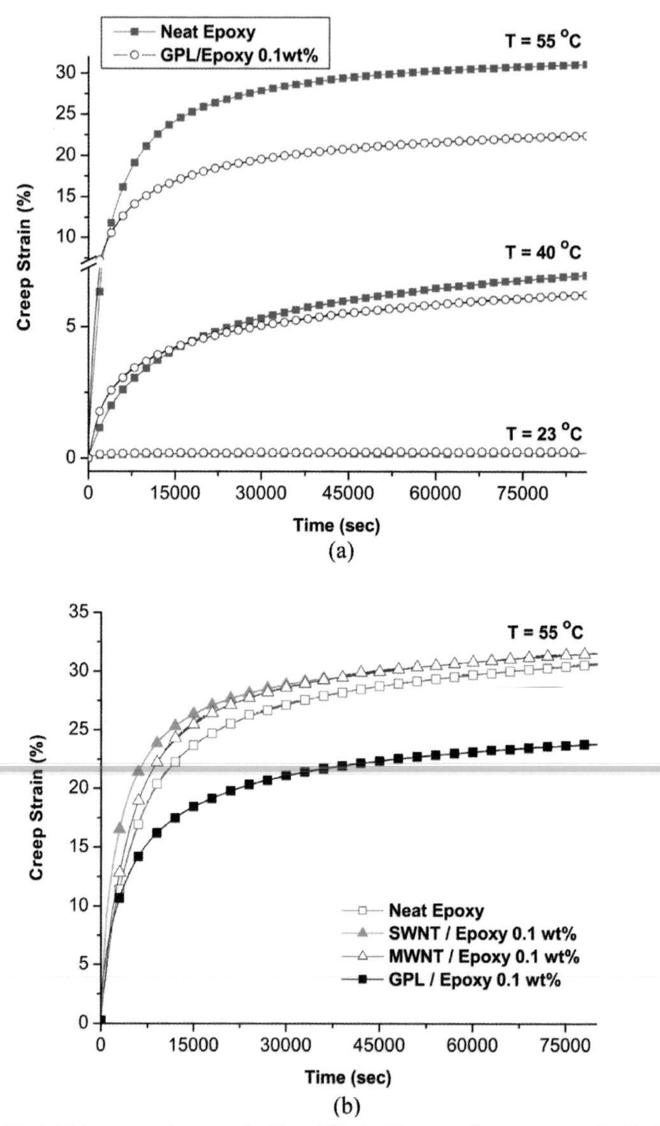

FIGURE 2.43. *(a) Macroscopic creep testing. Effect of temperature on creep for the baseline (unfilled epoxy) and the nanocomposite with ~0.1% of graphene additives. (b) Comparison of graphene vs. carbon nanotubes for creep reduction at elevated temperature. (Adapted from [57] with permission).*

epoxy-GPL creeps significantly less, which appears to be a consequence of the stronger interfacial adhesion between GPL and epoxy.

To conclude, addition of 0.1% weight fraction of GPL nanofillers to epoxy seems to significantly improve its creep behavior, both locally and at macro-scale. It appears this is an artifact of their high interfacial contact area and strong interfacial adhesion, which limits the mobility of the epoxy chains. Furthermore, the increase of the difference between the neat polymer response and that of the nanocomposites with increasing temperature indicates that thermally activated processes that are active in the neat epoxy are inhibited by the presence of the GPL. Graphene fillers, therefore, show good potential for creep suppression in polymers. However, improved dispersion of GPL fillers at higher weight fractions and testing with different polymer systems over a range of applied stresses and temperatures is necessary to evaluate the full impact of GPL as a creep-suppressant additive in polymers.

2.12. ELECTRICAL CONDUCTIVITY

Graphene-based additives can be highly effective in increasing the electrical conductivity of polymeric materials. This section discusses how graphene, as well as graphene oxide nanofillers in low concentrations, can be used to impart electrical conductivity to an otherwise insulating polyamide 6 (PA 6) polymer matrix [58]. The synthetic strategy employed to prepare electrically conducting PA 6 nanocomposites via one-step *in situ* polymerization and thermal reduction is summarized in Figure 2.44(a). Hydrophilic graphite oxide was first exfoliated to monolayer or few-layer nanosheets in water via ultrasonication, leading to a stable brown solution of exfoliated graphene oxide nanosheets. After ε-caprolactam monomer and 6-aminocaproic acid initiator were added, the reactants were further ultrasonicated. Under the protection of nitrogen and vigorous stirring, pre-polymerization and polymerization were conducted at 220°C and 260°C, respectively, to obtain PA 6/graphene oxide nanocomposites. Heating the suspension of graphene oxide in propylene carbonate at 150°C significantly reduced the graphene oxide nanosheets. Under the polymerization temperature, the graphene oxide is believed to be thermally reduced and its conductivity is thus expected to be greatly improved.

The results for electrical conductivity versus original graphene oxide content for the PA 6 nanocomposites is shown in Figure 2.44(b). For comparison, the curve of the PA 6 nanocomposites with conventional graphene (thermally exfoliated and reduced at ~1050°C from graphite oxide) is also plotted. It is seen that the introduction of graphene and graphene oxide significantly improved the electrical conductivity of PA 6 with a sharp transition from insulator to semiconductor. The inset of

Figure 2.44(b) indicates that the electrical conductivity (σ) of the nanocomposites obeys the power law:

$$\sigma \propto \sigma_f (\varphi - \varphi_c)^v \quad (57)$$

where σ_f, φ, and v are filler conductivity, filler volume fraction, and the critical exponent describing the rapid variation of σ near percolation threshold (φ_c), respectively. As shown in the inset of Figure 2.44(b), for the double-logarithmic plot of electrical conductivity versus ($\varphi - \varphi_c$), the conductivity of PA 6/graphene oxide nanocomposites agrees with the percolation behavior described by Equation (57).

It is known that graphene oxide and polyamide 6 are electrically insulating. However, during *in situ* polymerization, we expect that graphene oxide was partially reduced, thereby imparting electrical conductivity to the PA 6/graphene oxide nanocomposite. Percolation in PA 6/graphene oxide nanocomposites occurs when the filler concentration, φ_c, is near 0.41 vol. %. As the graphene oxide volume fraction increases from ~0.27 to ~1.09 vol. %, the electrical conductivity quickly rises by 10 orders of magnitude from ~4.2×10^{-14} to ~1.0×10^{-4} S/m. With only ~1.64 vol. % graphene oxide nanofiller, the conductivity approached ~0.028 S/m. These results suggest that the electrically insulating graphene oxide underwent thermal reduction during the polymerization from ε-caprolactam monomer to PA 6. Note that the graphene oxide nanocomposites are less conductive than the graphene nanocomposites. The later showed a lower percolation threshold of 0.22 vol. % and higher electrical conductivity of 1.1 S/m at the graphene content of 1.56 vol. %. This is not surprising and the difference can be attributed to their different extents of thermal reduction. Graphene was prepared by thermal reduction and exfoliation of graphite oxide at 1050°C, while the graphene oxide nanosheets were thermally reduced at a much lower temperature of 260°C during the *in situ* polymerization of ε-caprolactam.

To examine the composition variations of graphene oxide during the *in situ* polymerization, four samples (Figure 2.45) were selected for thermo-gravimetric analysis (TGA): (1) graphite oxide, (2) graphite oxide thermally reduced at 250°C for 10 h, (3) graphene oxide extracted from its PA 6 nanocomposite with formic acid, (4) neat polyamide 6. The graphite oxide exhibited a significant weight loss (about 80%) from 180°C to 260°C. However, the thermally reduced graphite oxide at 250°C had almost no weight loss under 260°C and less than 8% of weight loss in the whole process, indicating that thermal treatment at

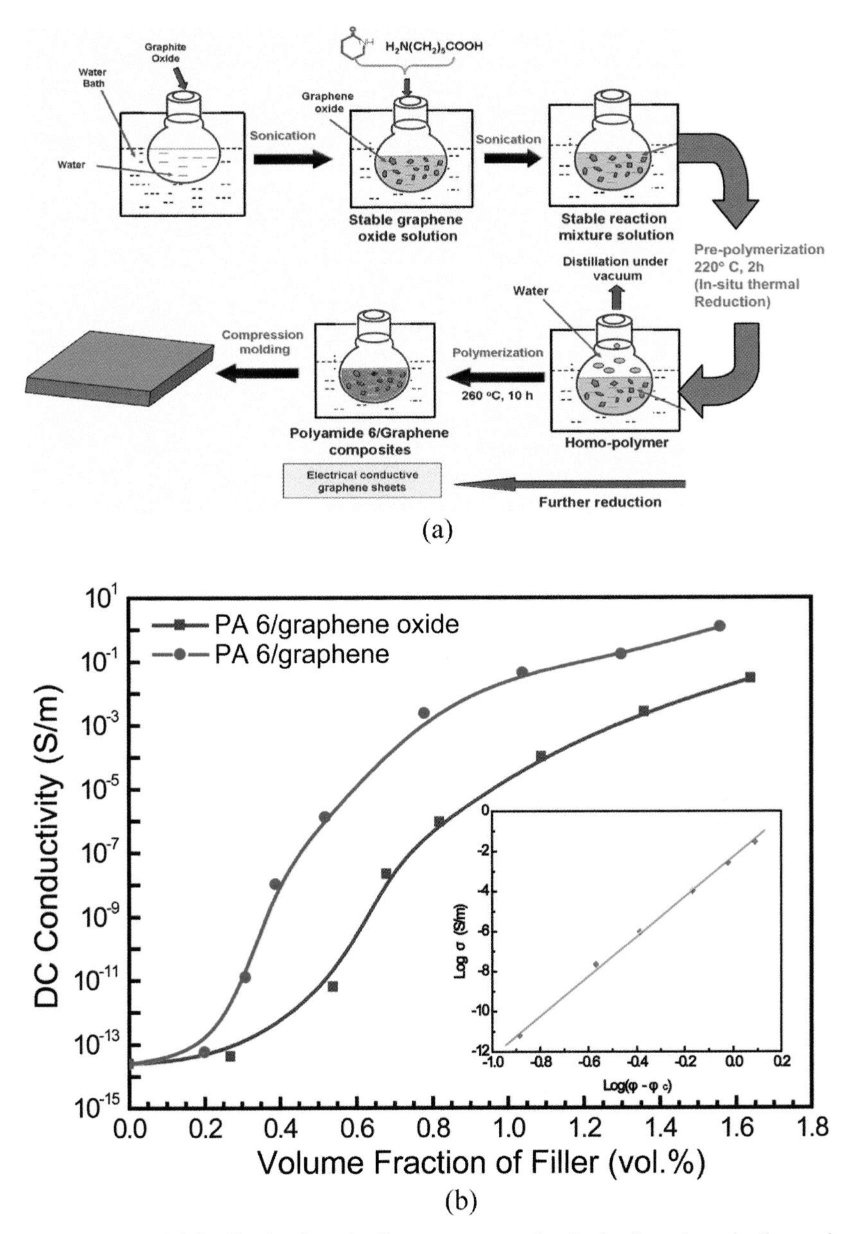

FIGURE 2.44. *(a) Synthesis of conductive nanocomposites by in situ polymerization and thermal reduction from graphene oxide. (b) DC electrical conductivity versus filler content for polyamide 6 (PA 6)/graphene oxide and PA 6/graphene nanocomposites. The inset is double-logarithmic plot of volume electrical conductivity versus ($\varphi - \varphi_c$) for PA 6/graphene oxide nanocomposites. (Adapted from [58] with permission).*

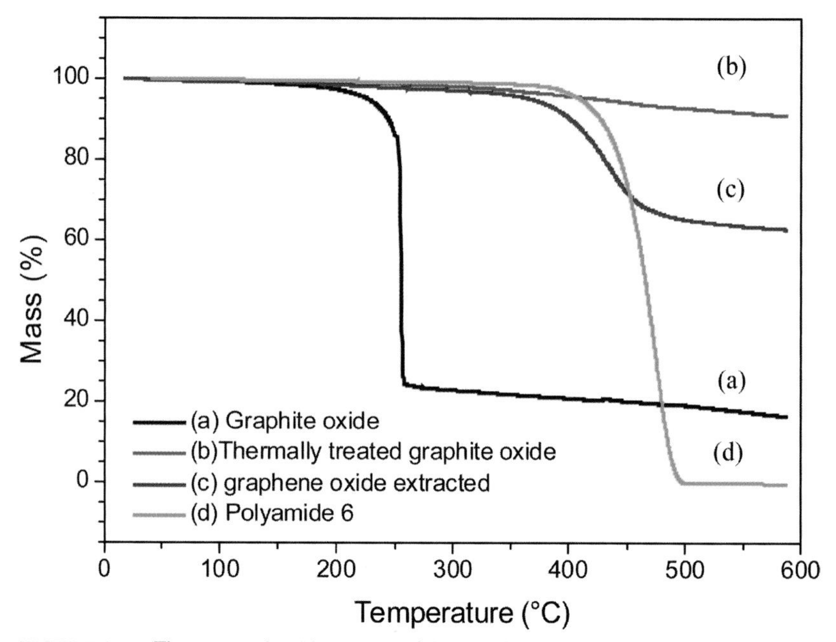

FIGURE 2.45. Thermogravimetric curves of (a) graphite oxide, (b) graphite oxide after thermal treatment at 250°C for 10 h, (c) graphene oxide extracted from its polyamide 6 nanocomposite using formic acid, and (d) neat polyamide 6. (Adapted from [58] with permission).

250°C is sufficient to eliminate most oxygen-containing groups on graphene oxide. Therefore, it is reasonable to assume that, after staying at 260°C for 10 h, the graphene oxide nanosheets should be reduced to conductive graphene. To confirm this, Figure 2.45(c) shows the TGA curve of the nanosheets extracted from the PA 6/graphene oxide nanocomposite. Because formic acid dissolves PA 6 easily, the sample C was obtained by using formic acid to wash the PA 6/graphene oxide nanocomposite several times until the weight did not change after being completely dried. There is almost no weight loss under 260°C, further proving that the graphene oxide in PA 6/graphene oxide nanocomposite has been substantially reduced in situ during the polymerization process. Note that the gradual weight loss occurring in a range of 350–480°C corresponds to the TGA curve of PA 6, which demonstrates that the samples C still contains ~40% PA 6 that was strongly coupled with graphene oxide and was still not washed away after repeated formic acid extraction. This is consistent with a recent report that indicates a covalent bonding reaction between PA 6 and the graphene oxide during in situ polymerization.

In addition to the *in situ* thermal reduction, the low percolation threshold and superior electrical conductivity of the PA 6 nanocomposites are also attributed to the high aspect ratio of the graphene oxide nanosheets and their uniform dispersion in the matrix. X-ray diffraction (XRD) can be employed to evaluate the exfoliation and dispersion of the graphene oxide and graphene in PA 6 matrix [Figure 2.46(a)]. There are no diffraction peaks derived from graphite or graphite oxide in the PA 6/graphene oxide nanocomposite and PA 6/graphene nanocomposite, indicating the exfoliated feature of graphene oxide and graphene nanosheets in the matrix. Additionally, the XRD pattern of neat PA 6 indicates it is crystallized in α-form, evidenced by its two characteristic peaks at 20.0° and 23.9° associated with the R (200) and R (002/202) planes. Interestingly, the γ-form crystallite of PA 6, evidenced by a new peak at 21.4°, was also observed in the two nanocomposites in addition to the α-form. The presence of the γ-form was due to the constraint imposed by the nanosheets on crystallization of PA 6 macromolecular chains.

To directly evaluate the overall dispersion and exfoliation of graphene oxide nanosheets in PA 6, Figure 2.46(b) and (c) show the SEM and TEM images of the PA 6 nanocomposites with 2 wt. % graphene oxide, respectively. The SEM image of the freeze-fractured surface shows that graphene oxide nanosheets were exfoliated and dispersed well in the matrix and no large graphene aggregates were present, which is consistent with the XRD results. TEM imaging can also be used to directly evaluate the exfoliation extent of the graphene oxide nanosheets. As shown in Figure 2.46(c), graphene oxide nanosheets existed as monolayer or few-layer nanosheets. The exfoliation of the graphene oxide enables the formation of an interconnected three-dimensional conducting network throughout the matrix. Thus, the electrically conducting feature of the PA 6/graphene oxide nanocomposite with low percolation threshold can be attributed to the stabilized exfoliation and dispersion of the thermally reduced graphene oxide nanosheets in the PA 6 matrix. In addition, the dispersion quality of graphene sheets was also observed with SEM and TEM. Similar to graphene oxide, good dispersion quality of the exfoliated graphene nanosheets was confirmed with TEM characterization of the PA 6/2 wt. % graphene nanocomposite, and no large aggregates were observed from the SEM image of the freeze-fractured surface of the same nanocomposite. This is related to the fact that graphene, in spite of thermal reduction at ~1050°C for 30 s, still bears some residual polar (hydroxyl) groups on its surface, which

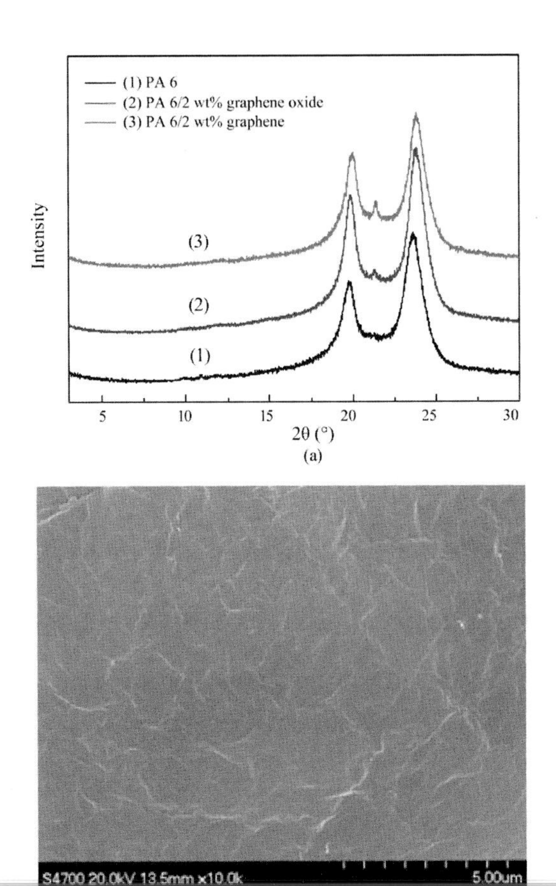

(b)

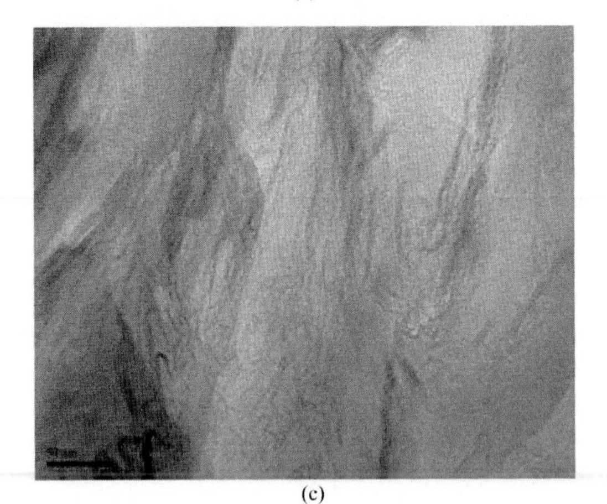

(c)

FIGURE 2.46. *(a) XRD patterns of (1) neat polyamide 6 (PA 6), (2) PA 6/2% graphene oxide nanocomposite, and (3) PA 6/2% graphene nanocomposite. (b) SEM microphotograph of freeze-fractured surface of PA 6/2 wt. % graphene oxide nanocomposite. (c) TEM microphotograph of PA 6/2 wt. % graphene oxide nanocomposite. The scale bar is 50 nm. (Adapted from [58] with permission).*

facilitates and stabilizes the dispersion of graphene nanosheets in the polar PA 6 matrix.

One of the interesting approaches that can be used to further enhance the electrical conductivity and lower the percolation threshold in graphene nanocomposites is to achieve selective localization of graphene in the polymer matrix. This effect can be demonstrated [59] in a polystyrene (PS)/graphene/polylactic acid (PLA) ternary (3-phase) composite system. Graphene/PS and multi-walled carbon nanotube (CNT)/PS composites were prepared by solution mixing followed by compression molding. Briefly, ~10 grams of PS was dissolved in ~60 ml dimethyl formamide (DMF) at ~45°C. The desired amounts of CNT or graphene were firstly dispersed in DMF (~0.1 g nanofiller per 100 ml DMF) by ultrasonic exfoliation for ~1.5 h using a KQ-250DB sonicator at 250 W. Then, the suspension of CNT or graphene was added into the PS solution. After the mixture was mechanically stirred for ~2 h, it was dropped into a large volume of vigorously stirred methanol to coagulate the PS nanocomposites, which were then filtered and dried in a circulating oven at ~80°C for ~16 h followed by drying in a vacuum oven at ~120°C for ~12 h. Finally, the dried nanocomposites were hot pressed into ~1-mm thick plates using a vacuum hot press at ~200°C and ~10 MPa. The same procedure is also used to prepare PS/PLA/graphene nanocomposites.

The volume conductivities of the nanocomposites are typically measured using a four-probe method. The advantage of this method is that the contact resistances are eliminated and the intrinsic resistivity of the graphene can be obtained. Figure 2.47(a) shows the electrical conductivities of the PS nanocomposites. It is clear that graphene is more effective in improving the electrical conductivity of PS than CNT, evidenced by higher electrical conductivity and smaller percolation threshold. The electrical conductivities of the PS/graphene nanocomposites show a rapid increase from ~6.7×10^{-14} to ~0.15 S/m, when the graphene content is increased from ~0.11 to ~0.69 vol. %. The addition of ~1.1 vol. % graphene endows PS with an electrical conductivity as high as ~3.49 S/m. However, for the CNT/PS nanocomposite with the same content of CNT (~0.69 vol.%), its conductivity is only ~3×10^{-5} S/m, which is nearly 4-orders of magnitude lower in comparison to the graphene/PS composites. As demonstrated by Xie *et al.*, compared to 1D CNT, 2D graphene nanosheets can form a conducting network at a lower content due to the latter's higher specific surface area, which is consistent with these results. In addition to this, the 2D sheet geometry of graphene should facilitate the formation of nanofiller-to-nanofiller contacts (i.e., a networked structure in the polymer) more easily than for 1D fillers such as carbon nanotubes.

Further, incorporation of PLA greatly improves the conductivity and reduces the percolation threshold of PS/graphene nanocomposites

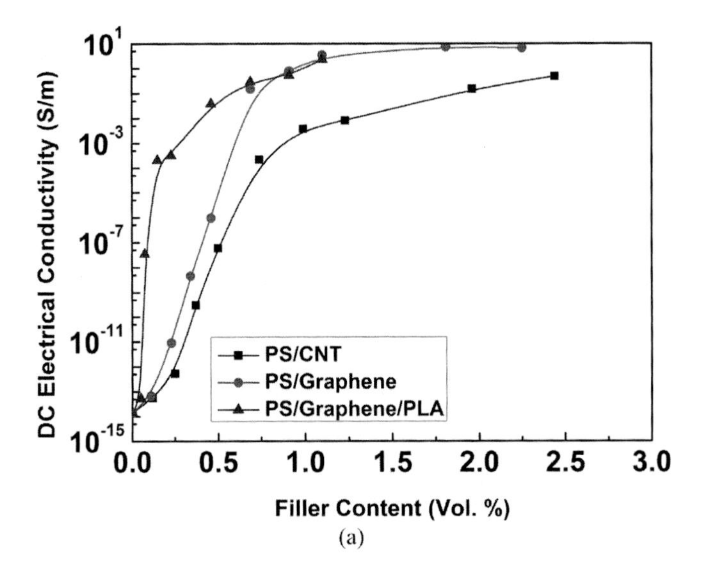

(a)

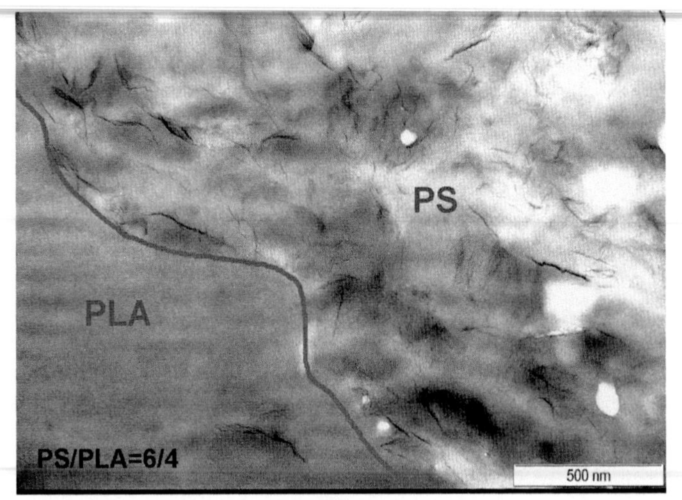

(b)

FIGURE 2.47. *(a) Electrical conductivity versus filler content for neat PS and its nanocomposites. (b) TEM image of PS/PLA (6/4) composites with ~0.46 vol. % (~1.0 wt. %) graphene additives. 6/4 indicates ~60% PS and ~40% PLA. The selective localization of graphene in the PS region is evident from the image. (Adapted from [59] with permission).*

[Figure 2.47(a)]. PS/PLA composites with ~60% PS and ~40% PLA were used in the testing. With 0.15 vol.% of graphene, the conductivity of the ternary graphene/PS/PLA composite reaches ~2.05×10^{-4} S/m. To achieve the same value in the PS/graphene binary composite, ~0.57 vol.% of graphene is required. The percolation threshold of the graphene/PS/PLA ternary composites is ~0.075 vol.%, which is ~4.5-fold lower than the PS/graphene binary composite. This result can be explained by selective localization of graphene in the PS/PLA components. Transmission electron microscopy (TEM) image of the composite [Figure 2.47(b)] indicate that the graphene nanosheets are selectively located in PS matrix, rather than in PLA phase. This selective localization of the graphene sheets results in a networked structure at relatively lower graphene content, which serves to significantly reduce the percolation threshold and increase the electrical conductivity of the nanocomposite.

2.13. THERMAL CONDUCTIVITY

Graphene exhibits a number of fascinating properties, including very high thermal conductivity. The value of thermal conductivity (κ) for single layer graphene is reported to be in the range of 4840–5300 W/mK [60], which is more than an order of magnitude higher than that of copper. This, coupled with its low density (~1–2 g/cm^3), makes graphene a very promising thermally conductive nanofiller in composite applications. Nanosheets of graphene have been used to improve thermal conductivity of different organic materials, such as epoxy, polypropylene, polystyrene, polyethylene, and polyamide [61–64]. This section describes thermal conductivity measurements performed [65] on 1-Octadecanol (stearyl alcohol), which is a well-known organic phase change material (PCM). PCMs can be used as latent heat storage and release units for thermal management of computers, electrical engines, and solar power plants, and for thermal protection of electronic devices. However, a major drawback of organic PCMs is their low thermal conductivity, which leads to large temperature gradients during heat transfer in or out of the material, reduced heat transfer rates, and large time constants. Therefore, increasing the thermal conductivity of the solidified material is one of the main issues in the application of organic PCMs. Here I will describe the effectiveness of graphene in boosting the thermal conductivity of 1-Octadecanol and its superiority as a thermally conductive nanofiller compared to carbon nanotubes and silver nanowires.

As mentioned above, 1-Octadecanol (stearyl alcohol) is an organic PCM, with a melting temperature of ~66°C, which is close to room temperature. It has an outstanding solid-liquid phase change enthalpy (~250 J/g). 1-Octadecanol is non-toxic, has a relatively low density (0.812 g/cm^3), and boils at ~210°C. To fabricate 1-Octadecanol/graphene composites, as seen in Figure 2.48, graphene sheets were first dispersed in a 25 ml/mg acetone solution and were ultrasonicated for 15 min (10 s on, 5 s off, 50% power). Then, the mixture was heated on a hot plate to 120°C while being sonicated for another 5 min. The 1-Octadecanol was then mixed with the graphene dispersion and was sonicated for another ~15 min on a hot plate at ~120°C. The mixture

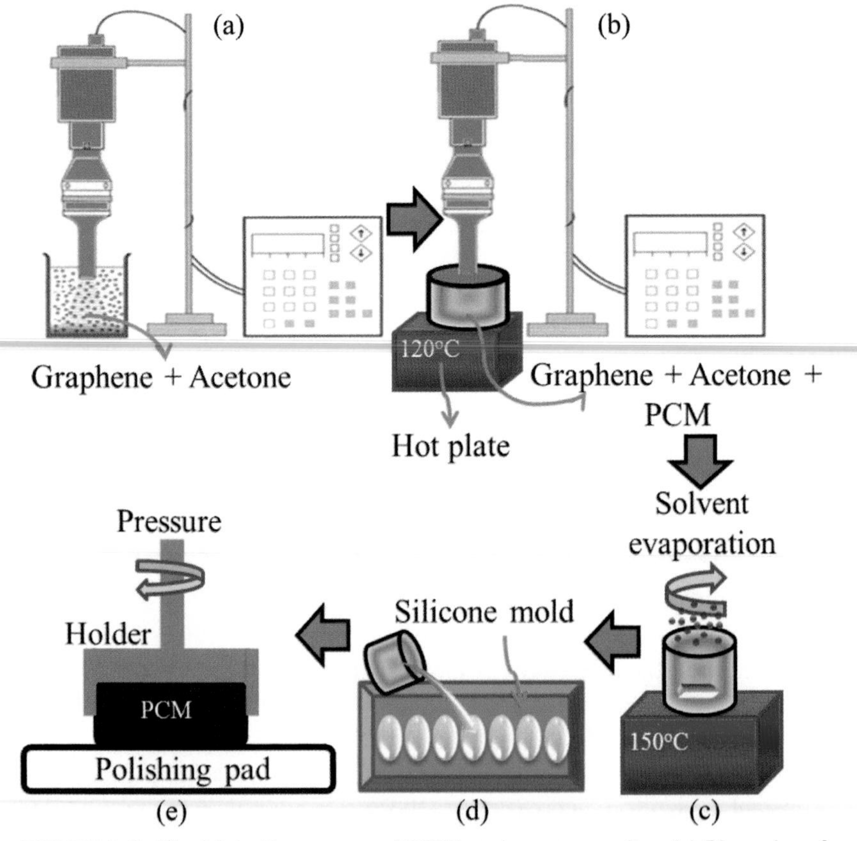

FIGURE 2.48. The fabrication process of PCM/graphene composites: (a) Dispersion of graphene in acetone using ultrasonic agitation. (b) Dispersion of graphene and PCM in acetone solution using heat and sonication. (c) Solvent evaporation using hot plate and stir magnet. (d) Insertion of the molten PCM in the mold. (e) Polishing the sample on sandpaper. (Adapted from [65] with permission).

of acetone, graphene, and PCM was stirred and heated to ~150°C to evaporate the remaining acetone. The nanocomposite in liquid phase was then poured into pre-heated Si rubber molds of cylindrical shape, ~6.35 mm thick and ~12.70 mm in diameter, and was left at room temperature to solidify for ~20 min. The resulting composite was affixed to a sample holder, cut to different thicknesses using a hot blade, and polished on a sandpaper pad.

The SEM images of the freeze-fractured surfaces of pristine 1-Octadecanol and 4% (by weight) graphene/1-Octadecanol nanocomposites are shown in Figure 2.49(a) and 2.49(b), respectively. High magnification images of 4% nanocomposite are also shown in Figure 2.49(c) and 2.49(d). As seen in these figures, graphene flakes were dispersed uniformly throughout the matrix, providing a three-dimensional network of high thermal conductivity graphene films. Good dispersion and network formation facilitate heat transfer and allow phonons to travel efficiently through the graphene fillers and between the flakes. It is well known that the type of the polymer matrix, degree of exfoliation of the graphene flakes, orientation of the fillers, and interfacial interaction influence the thermal transport in graphene nanocomposites. The SEM image of the fracture surface of an epoxy/graphene composite, fabricated using the same graphene flakes, is shown in Figure 2.49(e), which is comparable to Figure 2.49(d) in terms of magnification. However, there is a drastic difference in terms of the interaction of graphene with the surrounding polymer. In the case of graphene/1-Octadecanol, the graphene sheets are still covered with a thick layer of polymer, whereas in the case of graphene/epoxy, the graphene sheets seem to be completely separated from the epoxy matrix. This suggests there is a strong interface between graphene and 1-Octadecanol molecules.

To measure the heat conductivity of the composites, a steady-state 1D heat conduction method can be used. The experimental setup consists of an electrical heater, a heat sink and two thermocouples to measure the temperature gradient [Figure 2.50(a)]. To minimize the interface thermal resistance, fine-diameter electrically insulated thermocouples are embedded into two soft indium layers to measure the temperature at both sides of a thin cylindrical sample. Pressure is applied using a screw mechanism that is thermally insulated from the sample by a thick Teflon block. The heat losses in the experimental setup can be calibrated using glass samples of known conductivity. The thermal conductivity of the pure 1-Octadecanol measured with this setup (~0.38 W/mK) matches the value reported in the literature. To measure κ, the experimental ther-

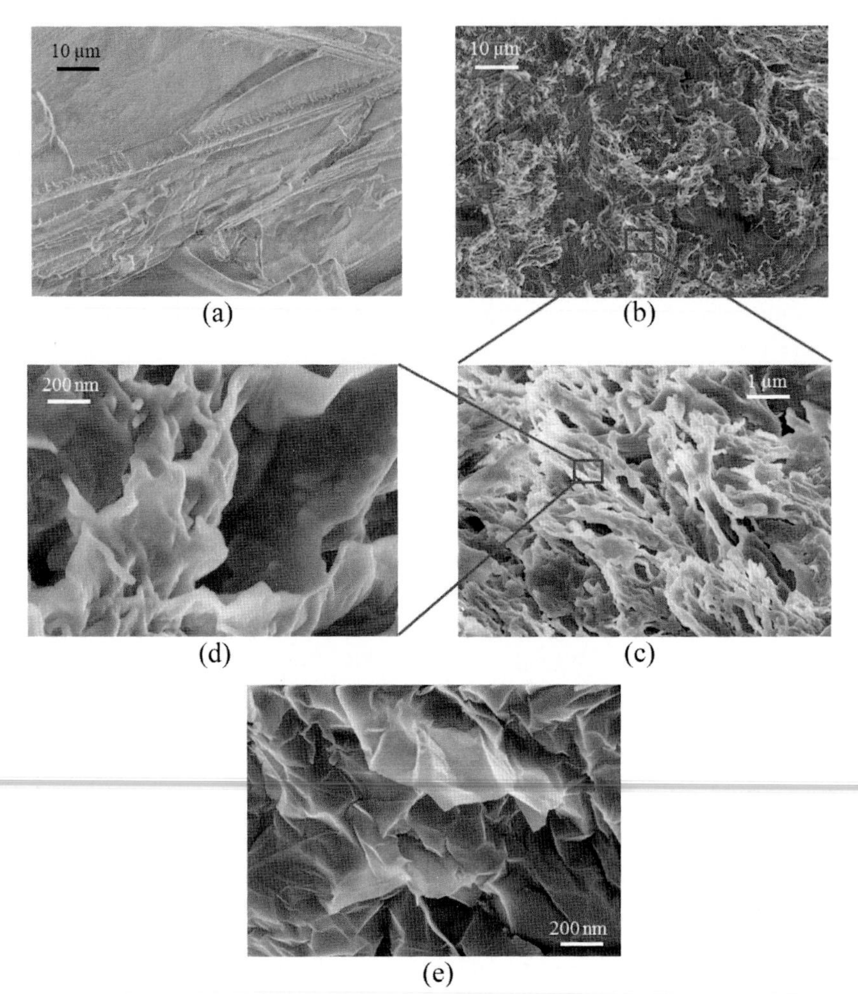

FIGURE 2.49. *Scanning electron microscopy (SEM) image of (a) fracture surface of pristine 1-Octadecanol. (b)–(d) Fracture surface of ~4% by weight graphene/1-Octadecanol composite with different magnifications. (e) Fracture surface of ~4% by weight graphene/epoxy composite. (Adapted from [65] with permission).*

mal resistance is first obtained from the slope of the temperature difference across the sample as a function of heater power [Figure 2.50(b)]. Next, the calibrated heat loss (R_{hl}) contribution is accounted for by using a parallel thermal resistance network model. To find the intrinsic thermal conductivity (κ), the interface thermal resistance (R_{int}) between the composite sample and the indium layer must be subtracted from the overall conduction resistance (R_t). This interface thermal resistance can be determined by testing samples with different thicknesses, then extrapolating the plot of $R_s + R_{\text{int}}$ vs. thickness to zero thickness using linear regression [Figure 2.50(c)]. The thermal resistance and conductivity of the sample are calculated using the following equations:

$$R_s = -R_{\text{int}} + \frac{R_t R_{hl}}{R_{hl} - R_t} \tag{58}$$

$$\kappa = \left(\frac{1}{R_s A}\right) \times t \tag{59}$$

where t and A are the thickness and cross-sectional area of the sample,

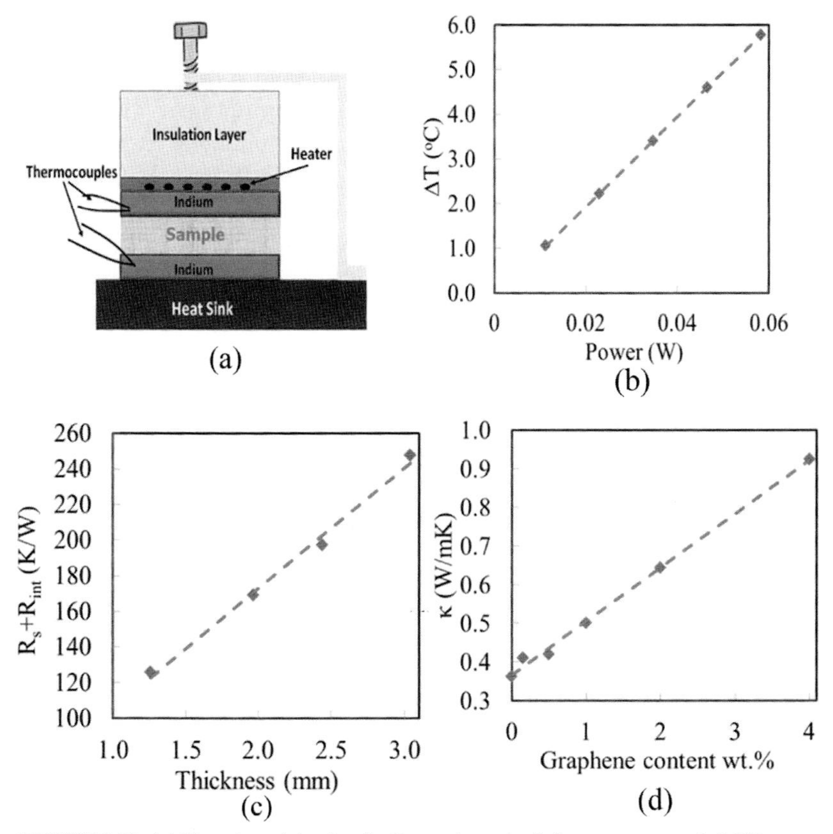

FIGURE 2.50. *(a) Experimental setup for thermal conductivity measurement. (b) Temperature difference vs. power for ~4% graphene/PCM nanocomposite. (c)* $R_s + R_{int}$ *(sample resistance + interface resistance) as a function of thickness to determine indium-sample thermal interface resistance. (d) Measured thermal conductivity of the sample vs. filler content for graphene/PCM composites. (Adapted from [65] with permission).*

respectively. The value of κ was obtained using the aforementioned method for the samples with different graphene contents. The results are shown in Figure 2.50(d). The thermal conductivity is considerably enhanced by the presence of graphene and reaches ~0.91 W/mK at ~4% graphene content. This value of κ is about 2.5-times higher than the measured κ value (~0.38 W/mK) for pure 1-Octadecanol. This increase can be attributed to the high thermal conductivity of the network of graphene fillers that provide a path of lower resistance for phonons to travel. Also, the high aspect ratio and large interfacial contact area of graphene, as well as the strong interface between graphene and the polymer, may help increase the thermal transport capacity of graphene nanocomposites.

Considering the very high theoretical conductivity of graphene (4840–5300 W/mK), one might expect a more dramatic improvement in thermal conductivity of the composite with the addition of graphene fillers. However, it has been reported that, since the dominant heat transfer mechanism is due to the lattice vibrations or phonons, poor phonon coupling in the vibrational modes at the polymer-filler and filler-filler interfaces cause thermal resistance, also called the Kapitza resistance. This resistance decreases the overall thermal conductivity of the material. Introduction of defects associated with the oxidation and thermal exfoliation of graphite can also reduce the thermal conductivity below the ideal value for defect-free graphene. Previous studies have also shown that the thermal conductivity enhancement due to the addition of conductive nanofillers differs depending on the type of nanofiller and the polymer matrix. For example, up to 4-fold increase in thermal conductivity can be attained by adding 5% GPLs (by weight) into epoxy, while only ~20% and ~26% increase in thermal conductivity was observed by adding the same amount of silver nanowires and multi-walled carbon nanotubes into PCMs, respectively. For PCM nanocomposites at 4% weight of graphene additives [see Figure 2.50(d)], the thermal conductivity is enhanced from ~0.38 W/mK to ~0.91 W/mK, an increase of ~140%. This indicates the superiority of graphene over multi-walled carbon nanotubes and silver nanowires as a thermally conductive nanofiller for PCM-based composites. There are two likely reasons for this:

(1) *Surface area:* For carbon nanotubes, only the outer surface of the tube is in contact with the matrix since the polymer chains are unable to penetrate into the inner pores of nanotubes. By contrast, for a flat sheet such as graphene, both surfaces of the sheet can

interface with the polymer, leading to greater interfacial contact area. The specific surface area (i.e., surface area per unit mass) of graphene is also markedly superior to solid silver nanowires.

(2) *Interface:* Graphene shows a rough and wrinkled (wavy) surface topology, which can enable strong interlocking and interfacial adhesion with the encapsulating polymer matrix. This could serve to reduce the interfacial thermal resistance at the nanofiller-matrix interface, leading to enhanced thermal conductivity.

2.14. GRAPHENE NANORIBBON-BASED COMPOSITES

So far we have considered GPLs as the nanofiller in polymer composites. An interesting variation of graphene is a graphene nanoribbon (GNR). GNRs are thin elongated strips of sp^2 bonded carbon atoms that can be fabricated by unzipping carbon nanotubes (CNTs). Pristine graphene is a semi-metal with zero electronic band gap. By contrast, a band gap can be opened in GNRs due to their more 1D structure. This has lead to great interest in the use of GNRs for semi-conductor device applications. GNRs have been produced by several techniques, including lithographic, chemical, sono-chemical, and chemical vapor deposition [66–68]. The solution-based oxidative method [69] is by far the most convenient method of synthesizing bulk quantities of GNRs for polymer composite applications. Multi-walled carbon nanotubes (MWNTs) can be unzipped based on a solution-based oxidative mechanism by engaging permanganate in an acid. A chemical reduction step is then typically used to relieve oxygen containing bonds resulting in graphene oxide nanoribbon (GONR) strips. The synthesized GONRs are then thermally reduced (heated to ~1050°C in ~35 sec) to expel oxygen groups and create GNRs. In this section, the processing and testing of the mechanical properties of GNR/epoxy composites is discussed. More specifically, this section presents the tensile strength, Young's modulus, ductility, and toughness of an epoxy polymer [70] reinforced with thermally treated GNRs. The results are compared to those of MWNT epoxy composites, in order to establish the effect of the unzipping role of the MWNT on the mechanical properties of the composite. The theoretically predicted elastic properties (using the Halpin-Tsai model) of GNR composites are also compared with the experimental data.

For GNR preparation, a mixture of sulfuric acid (98 %, 180 mL) and phosphoric acid (85.8 %, 20 mL) is added to multi-walled carbon nanotubes (e.g., Baytubes (Lot No. C70P, 1 g, 83 mmol)), and the mixture is stirred. Potassium permanganate (6 g, 38 mmol) is then added in 3 portions over ~30 min to the reaction mixture. After ~15 min, the mixture is heated to ~45°C and stirred at that temperature for ~24 h. The reaction mixture is then cooled down to room temperature and poured onto ice containing ~10 mL of hydrogen peroxide (30%). The GONRs are collected by centrifuging at ~4100 rpm for ~90 min. After the solution is decanted, the resulting wet GONRs can be re-dispersed in ~150 mL of hydrochloric acid (10%) and centrifuged again. This process is repeated at least two more times. Then, the wet GONRs are dispersed in ~50 mL de-ionized (DI) water and transferred to a dialysis bag and dialyzed in running DI water for ~1 week to remove the residual acid and inorganic salts. The water can then be removed under reduced pressure, and the GONRs are dried in a vacuum oven at ~70°C for ~16 h. Finally, the as-produced GONRs can be thermally reduced to GNRs by insertion for ~35 sec into a tube furnace, which is preheated to ~1050°C.

For epoxy nanocomposite processing, the GNRs are first dispersed in acetone (~200 mL of acetone per ~0.1 g of GNR) by high amplitude ultrasonication for ~1.5 h in an ice bath. A thermosetting epoxy resin (such as Epoxy-2000 from Fibreglast Inc., USA) can be added to the solution and sonicated, following the same procedure. The acetone is then removed through heating and magnet stirring the mixture for ~3 h at ~70°C. To eliminate any trace amount of acetone remaining, the mixture must be placed in a temperature-controlled vacuum chamber for ~12 h (at ~70°C). After, the GNR/epoxy blend cools down to room temperature, a low viscosity curing agent (e.g., 2120 from Fibreglast Inc., USA) can be added and blended using a high-speed shear mixer (e.g., ARE-250, Thinky, Japan) at ~2000 rpm (for ~4 min). Finally, the mixture is degassed for ~30 min in a vacuum chamber and is poured into silicon molds for curing under ~90 psi pressure (for 24 h) and post-curing at ~90°C (for 4 h). The same procedure can also be followed for dispersion of MWNTs in the epoxy matrix.

Figure 2.51(a)–(b) depicts scanning electron microscopy (SEM) images of MWNTs before and after the oxidation process. The purified MWNTs were procured from Bayer Corporation's Baytubes with average outer diameter of ~14 nm, inner diameter of ~10 nm, and length in the range of 1–10 μm. As seen in the SEM images, MWNTs were completely unzipped, resulting in complex, wavy-structured GONR strips. The typical width of the strips lies in the 50–100 nm range, while the length of the strips lies in the 1–10 μm range. The solution-based

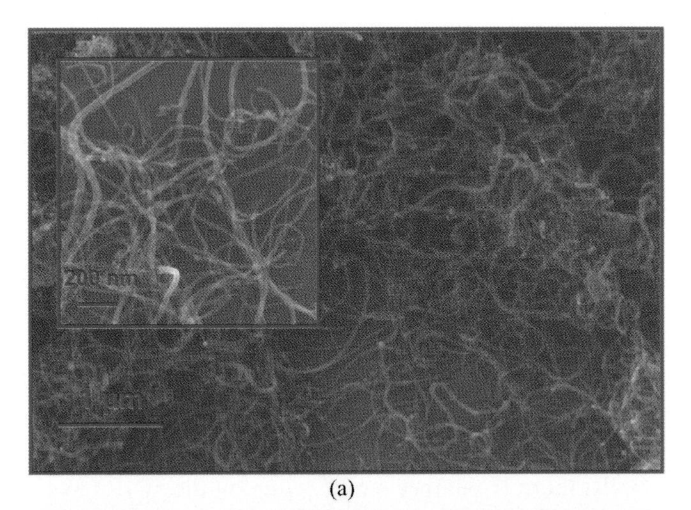

(a)

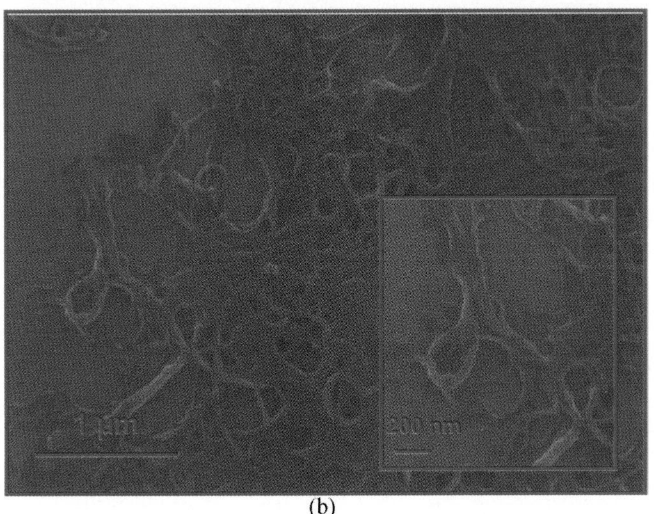

(b)

FIGURE 2.51. *Scanning electron microscopy (SEM) characterization of the multiwalled carbon nanotubes (a) before, and (b) after the oxidation process. Complete unzipping of the multi-walled carbon nanotubes into graphene oxide nanoribbons is evident from the images. As-produced graphene oxide nanoribbons were subsequently thermally reduced to yield graphene nanoribbons. (Adapted from [70] with permission).*

oxidative process generates a nearly 100% yield of nanoribbon structures by lengthwise cutting and unraveling of the MWNT side walls. Oxygen groups in the GONRs are eliminated by thermal reduction, as discussed above. The GNRs were uniformly dispersed by ultrasonication and high-speed shear mixing in the epoxy resin as described previously and cured to generate dog-bone-shaped coupons for the uniaxial tensile characterization. Figure 2.52(a) depicts a typical scanning electron micrograph of the freeze-fractured surface of the GNR/epoxy nanocomposite for ~0.3% weight fraction of GNR. The image shows GNR additives dispersed in the matrix; there was no indication of large agglomeration of clustering of the fillers. Moreover, the GNRs do not pull out easily from the fracture surface, as is typical of MWNTs. The inset in Figure 2.52(a) depicts a zoom-in SEM image of a GNR cluster that does appears to have been pulled out of the matrix. A low number density of such pulled out clusters was observed on the fracture surface of the composite. For the most part, the epoxy polymer appears to wet the GNR surface, which suggests a strong interface. Note that MWNTs suffer from intertube slip, which degrades their ability as structural reinforcement additives [71]. While interlayer slip is not completely overcome in GNR, there is a larger surface area on the GNRs available for interaction with the host epoxy.

Uniaxial tensile tests were conducted at a crosshead speed of ~1.5 mm/min at room temperature (~23°C) using an MTS-858 system operated in the displacement control mode. Figure 2.6(a) provides details regarding the sample geometry used for tensile testing. Figure 2.52(b) illustrates the typical stress-strain response of the baseline epoxy and GNR/epoxy composites for ~0.125%, ~0.3%, and ~0.4% weight fraction of GNR additives. The GNR composites exhibit significant increase in the Young's modulus and the ultimate tensile strength compared to the baseline epoxy at the expense of a drop in the ductility (i.e., the strain-to-break). The tests were also repeated for MWNT/epoxy composites at the same nanofiller loading fractions [Figure 2.52(c)]. The averaged results for the ultimate tensile strength (UTS) and Young's moduli (E) are summarized in Figure 2.53(a)–(b). The tensile strength of MWNT composites only increased by ~2–4% compared to the pristine epoxy, while the tensile strength of the ~0.3% weight GNR/epoxy nanocomposite increased by ~22%. There was no significant increase in the Young's modulus of the MWNT/epoxy composites, confirming that pristine MWNTs are poor additives for reinforcing highly cross-linked epoxy polymers that are unable to wrap around and interlock effectively with the atomistically smooth carbon nanotube surfaces. By contrast, for ~0.3% weight of GNR additives, the Young's modulus showed over 30% increase compared to the baseline epoxy and the MWNT-reinforced epoxy. Considering that these tests were performed on a heavily cross-linked epoxy (which displays an inherently high level of modulus/strength), this is an impressive level of increase in me-

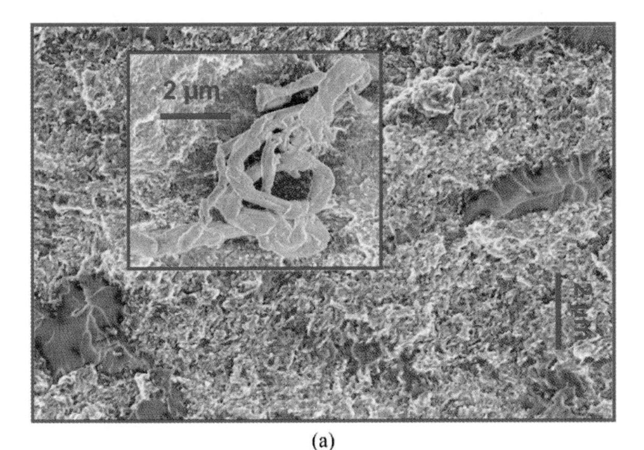

(a)

Stress (MPa)
80
70
60
50
40
30
20
10
0
0 0.01 0.02 0.03 0.04 0.05 0.06
Strain (mm/mm)
Epoxy
0.125% GNR, Epoxy
0.3% GNR, Epoxy
0.4% GNR, Epoxy

(b)

Stress (MPa)
70
60
50
40
30
20
10
0
0 0.01 0.02 0.03 0.04 0.05 0.06
Strain (mm/mm)
Epoxy
0.125% MWNT, Epo›
0.3% MWNT, Epoxy
0.4% MWNT, Epoxy

(c)

FIGURE 2.52. *(a) Typical SEM image of free-fractured surface of graphene nanoribbon epoxy composite with ~0.3% weight of nanofillers. Inset shows a high resolution image of an agglomerated nanoribbon cluster surrounded by the matrix-rich region. (b) Typical stress vs. strain curve of the baseline epoxy and nanocomposite formulations with varying weight fraction of graphene nanoribbon additives. (c) Corresponding stress vs. strain response for MWNT/epoxy composites. (Adapted from [70] with permission).*

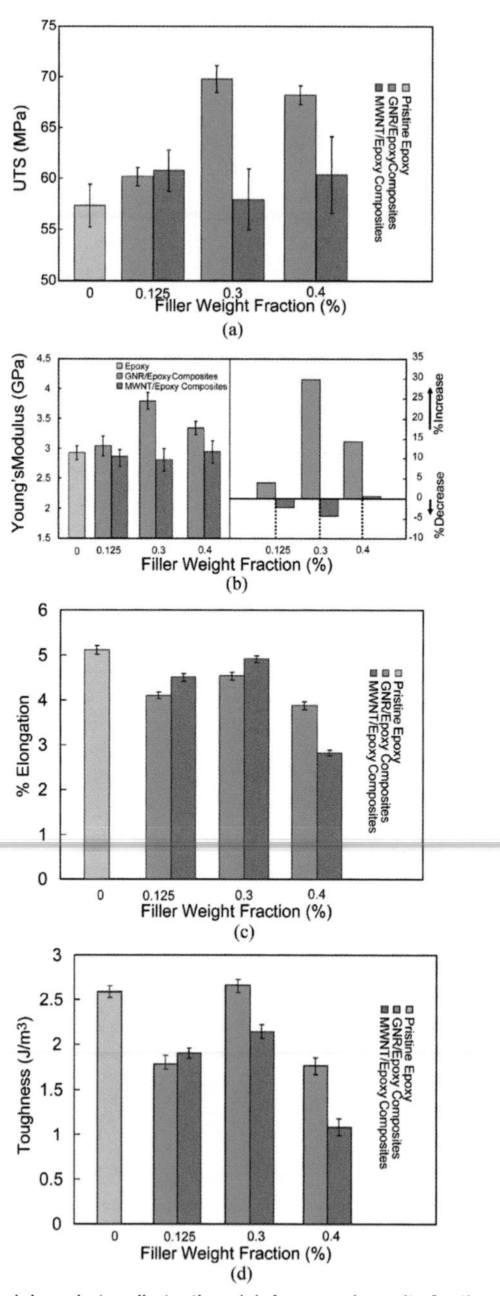

FIGURE 2.53. *Uniaxial mode tensile testing. (a) Averaged results for the ultimate tensile strength (UTS) for the baseline epoxy and nanocomposites with varying loading fraction of nanofillers. (b) Absolute value and percentage changes in the Young's modulus of the pristine epoxy and nanocomposites with various nanofiller weight fractions. (c) Percent elongation at failure for the baseline epoxy and nanocomposites with various nanofiller weight fractions. (d) Corresponding results for the material toughness. (Adapted from [70] with permission).*

chanical properties at low nanofiller loading. By comparison, for ~1% weight of functionalized single-walled carbon nanotube (SWNT) fillers in epoxy, researchers have reported ~30% and ~15% increases in the Young's modulus and the tensile strength, respectively. Similar levels of enhancement in the modulus and strength for GNR composites can be achieved at ~70% lower weight fraction of the nanofillers.

In addition to tensile strength and modulus, it is also important to compare the ductility (i.e., strain-to-break) and the toughness (i.e., energy absorbed at failure, which is the total area under the stress vs. strain curve) of the baseline and nanocomposite epoxies. Figure 2.53(c) indicates that the ductility of the GNR composites is ~10–15% lower than the pristine epoxy. The reduced ductility may be caused by stress concentration in the vicinity of the filler; this typically occurs when hard fillers are incorporated into a brittle matrix. Besides, agglomeration of GNR can lead to defects in the matrix that can act as seed points for crack initiation and premature fracture. Similar loss of ductility was also seen for MWNT composites [Figure 2.52(c) and 2.53(c)]. Ductility is usually critical for metals where the manufacturing processes are based on metal forming operations (e.g., rolling, extrusion). For epoxies, 10–15% loss in ductility can be tolerated, provided that other mechanical properties such as modulus and strength are enhanced. Figure 2.53(d) shows the material toughness for the baseline epoxy and the MWNT and GNR epoxy nanocomposites. MWNT composites show lower toughness than the baseline epoxy, while for the ~0.3% weight fraction GNR/epoxy composite, the toughness is marginally increased compared to the baseline epoxy.

To predict the elastic properties of the GNR/epoxy composites, the GNR strips were modeled as rectangular cross-section fibers having width (W), length (L), and thickness (t). The well-established Halpin-Tsai equations (Section 2.2) developed for randomly-oriented fiber reinforcement were applied as follows [70]:

$$\alpha = \frac{E_C}{E_M} = \frac{3}{8}\left(\frac{\zeta\eta_L V_{GNR}+1}{1-\zeta\eta_L V_{GNR}}\right)+\frac{5}{8}\left(\frac{2\zeta\eta_W V_{GNR}+1}{1-\zeta\eta_W V_{GNR}}\right) \tag{60}$$

$$\eta_L = \frac{\beta-1}{\beta+\zeta} \tag{61}$$

$$\eta_W = \frac{\beta-1}{\beta+2} \tag{62}$$

where α is the ratio of the composite's modulus (E_C) to that of the pure epoxy (E_M), E_{GNR} = 1 TPa is the Young's modulus of GNRs, V_{GNR} in the GNR volume fraction, and ζ and β are constants, which are defined as follows:

$$\zeta = \frac{W + L}{t} \tag{63}$$

$$\beta = \frac{E_{GNR}}{E_M} \tag{64}$$

Substituting from Equations (59)–(62) into Equation (58), the ratio of the composite's modulus (E_C) to that of the pure epoxy (E_M) can be expressed as a function of the volume fraction of reinforcements (V_f) as follows:

$$\alpha = \frac{E_C}{E_M} = \left[\frac{3}{8}\left(\frac{((W+L)/t)\dfrac{(E_{GNR}/E_M)-1}{(E_{GNR}/E_M)+((W+L)/t)}V_{GNR}+1}{1-\dfrac{(E_{GNR}/E_M)-1}{(E_{GNR}/E_M)+((W+L)/t)}V_{GNR}} \right) + \frac{5}{8}\left(\frac{2\dfrac{(E_{GNR}/E_M)-1}{(E_{GNR}/E_M)+2}V_{GNR}+1}{1-\dfrac{(E_{GNR}/E_M)-1}{(E_{GNR}/E_M)+2}V_{GNR}} \right) \right] \tag{65}$$

The weight fraction of GNRs is converted into volume fraction based on the estimated densities of GNRs and the epoxy. Based on information provided by the MWNT supplier and the SEM characterization, the average size of GNRs can be estimated as ~5 μm in length (L), ~78 nm in width (W), and ~4 nm in thickness (t). As a conservative estimate, the density of the GNRs can be taken as the standard density of graphite (~2.25 g/cm^3). Figure 2.54 shows the results of the theoretical predictions from Equation (65) in comparison to the experimental values. There is reasonable agreement between theory and experiment until ~0.13% volume fraction (i.e., 0.3% weight fraction), beyond which the experimental data for the Young's modulus begins to drop off below the theoretical prediction. This suggests that the quality of dispersion of GNR in the epoxy resin begins to degrade beyond a weight fraction of ~0.3%. This result is consistent with 2D GPLs, which begin to agglom-

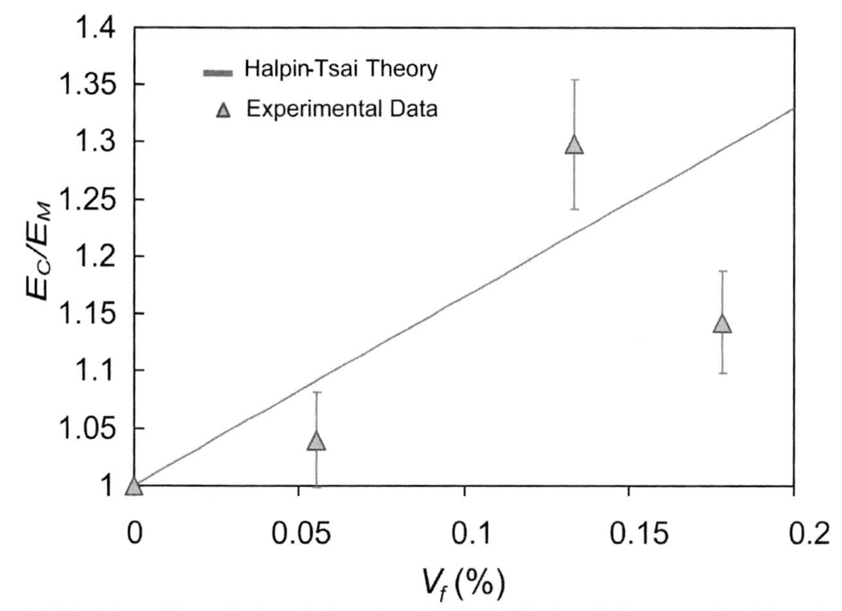

FIGURE 2.54. Theoretical predictions from the Halpin-Tsai model for normalized Young's modulus of graphene nanoribbon composites plotted versus nanofiller volume fraction, indicating reasonable agreement between theory and experimental data up to ~0.13% volume fraction (~0.3% weight fraction) of nanofillers. (Adapted from [70] with permission).

erate at even lower weight fractions (~0.1%) in epoxy matrices (Section 2.2). This highlights the need for continued research to develop new methods to enhance GNR dispersion at higher loading fractions in order to derive the full benefit of these unique materials for structural composites.

The surface chemistry and defect density of MWNTs, GONRs, and GNRs can be studied using techniques such as X-ray photoelectron spectroscopy (XPS) and Raman spectroscopy. The surface chemistry of the MWNTs and GNRs used in the testing is very similar as confirmed by XPS (Figure 2.55). The elimination of oxygen-containing groups was confirmed by the C1s spectrum of GNR, since no peaks corresponding C–O (~286 eV) and C=O (~287.8 eV) bonds were observed for the thermally reduced GNRs. Consequently, surface chemistry alone is not responsible for enhanced effectiveness of GNRs over MWNTs as a structural reinforcement additive. The Raman spectra for GNRs and MWNTs (Figure 2.56, Table 2.2) indicate that the ratio of the integrated intensity of the D band to G band is ~1.25 for GNRs compared to ~1.04 for MWNTs, indicating that the GNRs are some-

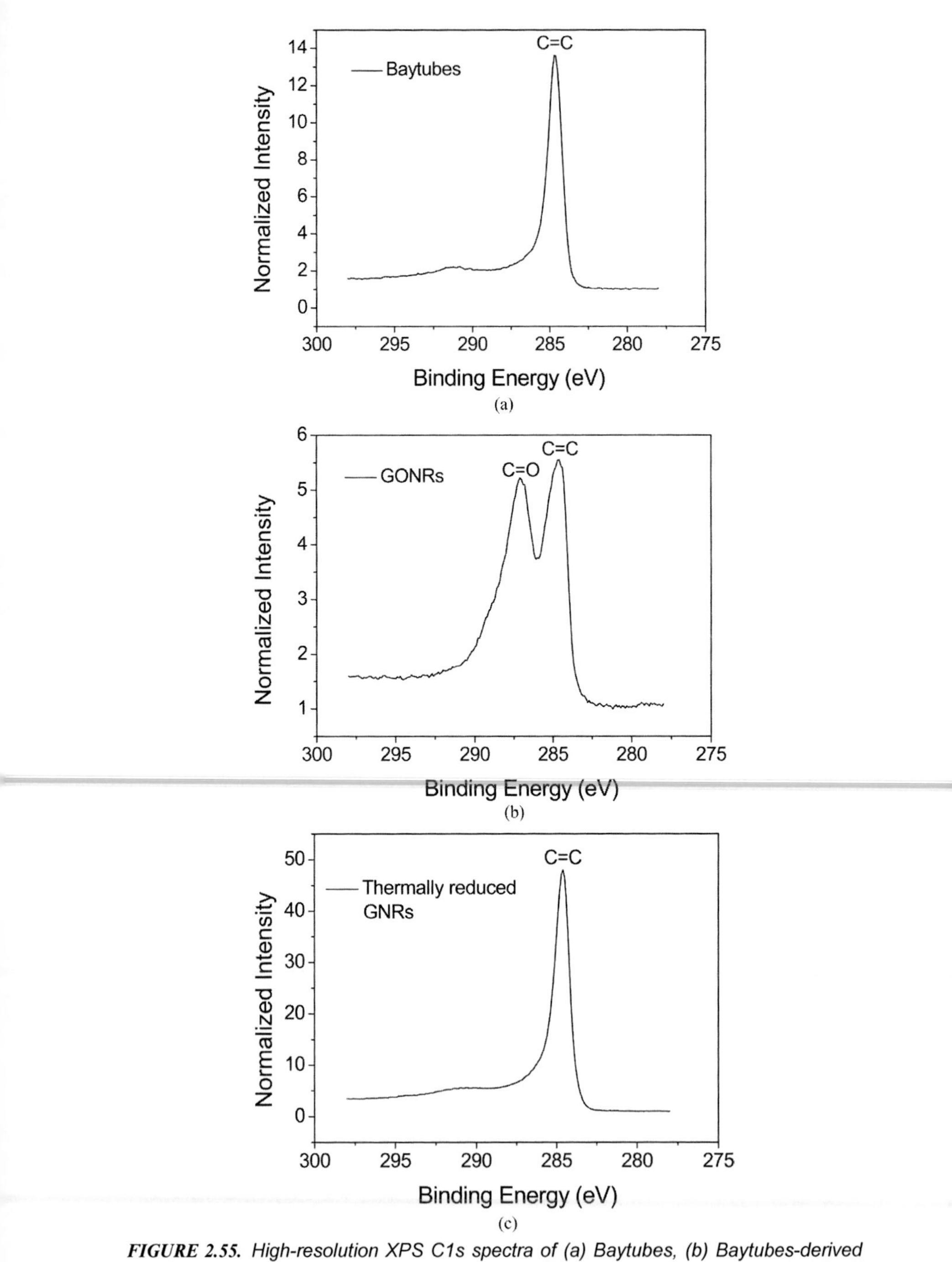

FIGURE 2.55. *High-resolution XPS C1s spectra of (a) Baytubes, (b) Baytubes-derived GONRs, and (c) the thermally reduced GNRs. Signals at ~284.8 eV and ~287.8 eV correspond to C=C and C=O, respectively. Elimination of oxygen is confirmed by the C1s spectrum of thermally reduced GNRs, since no peaks for C–O (~286 eV) and C=O (~287.8 eV) bonds were observed. (Adapted from [70] with permission).*

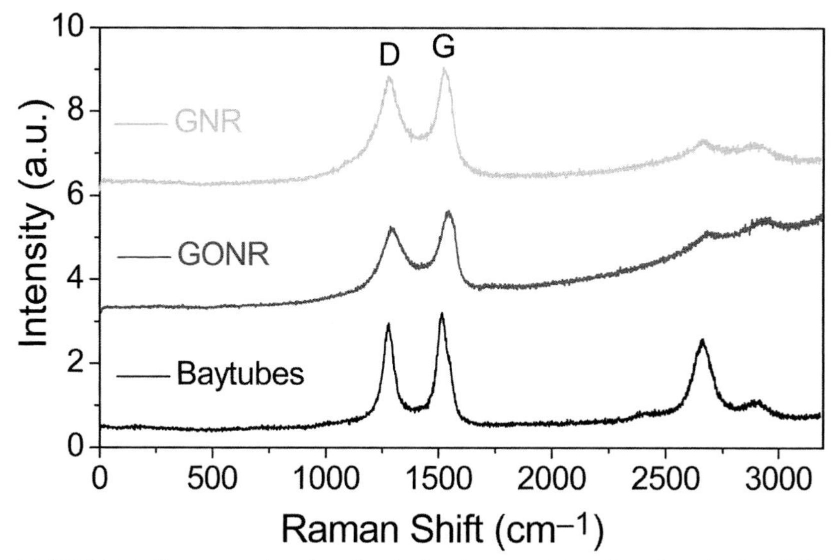

FIGURE 2.56. *Raman spectra of multi-walled carbon nanotubes (from Baytubes), GONR, and thermally reduced GNR are shown. From the Raman spectra and the ratio of the integrated intensities of the D band to G band, it is clear that the nanotubes (Baytubes) have substantial defects. This is a result of the Bayer Corporation synthesis process for their large-scale route to these compounds (250 tons per year). The intensity ratio of the D and G bands increases from ~1.04 for Baytubes to ~1.25 for GNR. (Adapted from [70] with permission).*

what more defective than MWNTs, and likely provide greater handles for interaction with the host epoxy. More importantly, however, is the surface area of the nanoribbons relative to the tubular structures; therefore, taken together, the nanoribbons are superior. The average specific surface area of the MWNT and GNR samples in the powder form was measured by standard BET N_2 cryo-sorption experiments. The specific surface area of GNRs (~511 m^2/g) was found to be significantly greater than MWNTs (~291 m^2/g). This confirms that the combined GNR surface areas are far greater than the MWNTs from which they are derived.

TABLE 2.2. The Integrated Intensities and the Intensity Ratio (I_D/I_G) of the D and G Bands for the MWNTs (from Baytubes), Graphene Oxide Nanoribbons (GONRs), and Thermally Reduced Graphene Nanoribbons (GNRs). (Adapted from [70] with permission).

	D band (Integrated Area: ID)	G band (Integrated Area: IG)	ID/IG
Baytubes	107546.35	103402.92	1.04
GONR	304696.53	280172.23	1.09
GNR	218901.10	175156.09	1.25

This likely contributes considerably to the strong interfacial interaction of the GNRs in epoxy rather than MWNT interactions in epoxies at the same weight loadings.

In summary, while pristine MWNTs are ineffective at reinforcing epoxy composites, unzipping them into GNRs results in significant improvement. The two main factors responsible for this are:

(1) *Surface Area:* Unraveling of MWNTs into GNR platelets generates a significant increase in the interfacial contact area since both surfaces of each individual GNR platelet will contact the matrix, as opposed to only the outermost cylinder of the MWNTs.

(2) *Geometry:* It is challenging for highly cross-linked polymers such as epoxies to wrap around tubular MWNTs (with 10–20 nm diameter) and mechanically interlock with them. We expect that it is easier for such polymers to adhere to a flat nanofiller with a sheet or ribbon-like geometry. The GNRs are also more defective than MWNTs, which contributes to better interfacial binding. These results indicate that GNR shows significant potential as a structural reinforcement additive in polymer-based composite materials.

2.15. REFERENCES

1. Rafiee, J.; Rafiee, M. A.; Yu, Z.; Koratkar, N. Super-hydrophobic to Superhydrophilic wetting control in graphene films. *Adv. Mater.* **2010**, *22*, 2151–2154.
2. Schniepp, H. C., Li J.-L.; McAllister, M. J.; Sai, H.; Herrera-Alonso, M.; Adamson, D. H; Prud'homme, R. K.; Car, R.; Saville, D. A.; Aksay, I. A. Functionalized single graphene sheets derived from splitting graphite oxide. *J. Phys. Chem. B* **2006**, *110*, 8535–8539.
3. McAllister, M. J.; Li, J.-L.; Adamson, D. H.; Schniepp, H. C.; Abdala, A. A.; Liu, J.; Herrera-Alonso, M.; Milius, D. L.; Car, R.; Prud'homme, R. K.; Aksay, I. A. Single sheet functionalized graphene by oxidation and thermal expansion of graphite. *Chem. Mater.* **2007**, *19*, 4396–4404.
4. Pilato, L.A.; Michno, M. J. Advanced Composite Materials. 1st ed.; Berlin: Springer, 1994.
5. Garg, A.C.; Y.-W. Mai. Failure mechanisms in toughened epoxy resins—A review. *Compo. Sc. Techno.* **1988**, *31*, 179–223.
6. Park, S.; Ruoff, R. S. Chemical methods for the production of graphenes. *Nat. Nanotech* **2009**, *4*, 217–224.
7. Rao, C. N. R.; Sood, A. K.; Subrahmanyam, K. S.; Govindara, A. J. Graphene: The new two-dimensional nanomaterial. *Agewandte Chemie* **2009**, *48*, 7752–7777.
8. Soldano, C.; Mahmood, A.; Dujardin, E. Production, properties and potential of grapheme. *Carbon* **2010**, *48*, 2127–2150.
9. Inagaki, M.; Kim, Y. A.; Endo, M. Graphene: preparation and structural perfection. *J. Mater. Chem.* **2010**, *21*, 3280–3294.
10. Ramanathan, T.; Abdala, A. A.; Stankovich, S.; Dikin, D. A.; Herrera-Alonso, M.; Piner, R. D.; Adamson, D. H.; Schniepp, H. C.; Chen, X.; Ruoff, R. S.; Nguyen, S. T.; Aksay, I. A.; Prud'Homme, R. K.; Brinson, L. C. Functionalized graphene sheets for polymer nanocomposites. *Nat. Nanotechnol.* **2008**, *3*, 327–331.

11. Halpin, J. C.; Kardos, J. L. The Halpin-Tsai equations: a review. *Polym. Eng. Sci.* **1976**, *16*, 344–352.
12. Johnsen, B.B., et al. Toughening mechanisms of nanoparticle-modified epoxy polymers. *Polymer* **2007**, *48*(2), 530–541.
13. Lewis, T. B.; Nielsen, L. E. Dynamic mechanical properties of particulate-filled composites. *J. Appl. Polym. Sci.* **1970**, *14*(6), 1449–1471.
14. McGee, S.; McCullough, R. L. Combining rules for predicting the thermoelastic properties of particulate filled polymers, polymers, polyblends, and foams. *Polym. Composite* **1981**, *2*(4), 149–161.
15. Nielsen, L.E.; Landel, R.F. Mechanical Properties of Polymers and Composites. 2nd.; CRC Press, 1994.
16. Rafiee, M. A.; Rafiee, J.; Srivastava, I.; Wang, Z.; Song, H.; Yu, Z.; Koratkar, N. Fracture and fatigue in graphene nanocomposites. *Small* **2010**, *6*, 179–183.
17. Rafiee, M. A.; Rafiee, J.; Yu, Z.-Z.; Koratkar, N. Buckling resistant graphene nanocomposites. *Appl. Phys. Lett.* **2009**, *95*, 223103.
18. Rafiee, M. A.; Rafiee, J.; Wang, Z.; Song, H.; Yu Z.-Z.; Koratkar, N. Enhanced mechanical properties of nanocomposites at low graphene content. *ACS Nano* **2009**, *3*, 3884–3890.
19. Schadler, L. S.; Giannaris, S. C.; Ajayan, P. M. Load transfer in carbon nanotube epoxy composites. *Appl. Phys. Lett.* **1998**, *98*, 252.
20. Orowan, E. Fracture and strength of solids. *Rep. Prog. Phys.* **1949**, *185*.
21. Zhang, W.; Srivastava, I.; Zhu, Y.-F.; Picu, C. R.; Koratkar, N. Heterogeneity in Epoxy Nanocomposites Initiates Crazing: Significant Improvements in Fatigue Resistance and Toughening. *Small* **2009**, *5*, 1403–1407.
22. Koratkar, N.; Suhr, J.; Joshi, A.; Kane, R.; Schadler, L.; Ajayan, P.; Bartolucci, S. Characterizing energy dissipation in single-walled carbon nanotube polycarbonate composites. *Appl. Phys. Lett.* **2005**, *87*, 063102
23. Blackman, B. R. K.; Kinloch, A. J.; Sohn Lee, J.; Taylor, A. C.; Agarwal, R.; Schueneman, G.; Sprenger, S. The fracture and fatigue behaviour of nano-modified epoxy polymers. *J. Mater. Sci.* **2007**, 42, 7049–7051.
24. Wetzel, B.; Rosso, P.; Haupert, F.; Friedrich, K.. Epoxy nanocomposites—fracture and toughening mechanisms. *Eng. Fract. Mech.* **2006**, *73*, 2375–2398.
25. Rafiee, M.A.; Rafiee, J.; Yavari, F.; Koratkar, N. Fullerene/Epoxy Nanocomposites- Enhanced Mechanical Properties at Low Nanofiller Loading. *J. Nanopart. Res.* **2011**, *13*, 733–737
26. Zerda, A. S.; A. Lesser, J. Intercalated Clay Nanocomposites: Morphology, Mechanics, and Fracture Behavior. *J. Polym. Sci. Pol. Phys.* **2001**, *39*, 1137–1146.
27. Wang, K.; Chen, L.; Wu, J.; Toh, M. L.; He, C.; Yee, A. F.. Epoxy nanocomposites with highly exfoliated clay: mechanical properties and fracture mechanisms. *Macromolecules* **2005**, *38*, 788–800.
28. Liu, W.; Hoa, S. V.; Pugh, M. Fracture toughness and water uptake of high-performance epoxy/nanoclay nanocomposites. *Compos. Sci. Technol.* **2005**, *65*, 2364–2373.
29. Gojny, F. H.; Wichmann, M. H. G.; Fiedler, B. K.; Schulte, K. Influence of different carbon nanotubes on the mechanical properties of epoxy matrix composites—A comparative study. *Comp. Sci. Technol.* **2005**, *65*, 2300–2313.
30. Paris, P.C.; M.P. Gomez; W.P. Anderson, A Rational Analytic theory of fatigue. *The Trend in Engineering.* **1961**, *13*, 9–14.
31. Mai, Y.W.; J.G. Williams. Temperature and enviromental effects on the fatigue fracture in polystyrene. *J. Mater. Sci.* **1979**, *14*(8), 1933–1940.
32. E 647 Standard Test Method for Measurement of Fatigue Crack Growth Rates. American Society for Testing and Materials 1997.
33. Zhang, W.; Picu, C. R.; Koratkar, N. Suppression of fatigue crack growth in carbon nanotube composites. *Appl. Phys. Lett.* **2007**, *91*, 193109.

34. Zhang, W.; Picu, C. R.; Koratkar, N. The effect of carbon nanotube dimensions and dispersion on the fatigue behavior of epoxy nanocomposites. *Nanotechnology* **2008**, *19*, 285709.
35. Faber, K.T.; A.G. Evans. Crack deflection processes. I—Theory. *Acta Metallurgica* **1983**, *31*, 565–576.
36. Lange, F.F. Interaction of a crack front with a second-phase dispersion. *Philo. Mag.* **1970**, *22*(179), 983–992.
37. Zhao, S. Mechanical and Thermal properties of nanoparticle filled epoxy. Nanocomposites in Materials Science and Engineering; Rensselaer Polytechnic Institute: Troy, 2007; p 220.
38. Huang, Y., et al., Mechanisms of Toughening Thermoset Resins. *Adv. in Chem.* **1993**, *233*, 1–35.
39. Azimi, H.R., Pearson, R.A.; Hertzberg, R.W. Fatigue of hybrid epoxy composites: Epoxies containing rubber and hollow glass spheres. *Polym. Eng. Sci.* **1996**, *36*, 2352–2365.
40. Kinloch, A.J.; Williams, J.G.. Crack blunting mechanisms in polymers. *J. Mater. Sci.* **1980**, *15*(4), 987–996.
41. Suhr, J.; Koratkar, N.; Keblinski, P.; Ajayan P. Viscoelasticity in carbon nanotube composites. *Nat. Mater.* **2005**, *4*, 134–137.
42. Starr, F. W.; Schroder, T. B.; Glotzer, S. C. Molecular dynamics simulation of a polymer melt with a nanoscopic particle. *Macromolecules* **2002**, *35*, 4481–4492.
43. Srivastava, 0. I.; Mehta, R. J.; Yu, Z.-Z.; Schadler, L.; Koratkar, N. Raman Study of Interfacial Load Transfer in Graphene Nanocomposites. *Appl. Phy. Lett.* **2011**, *98*, 063102.
44. Ni, Z.H., et al. Uniaxial Strain on Graphene: Raman Spectroscopy Study and Band-Gap Opening. *ACS Nano* **2008**, *2*(11), 2301–2305.
45. Cooper, C.A.; Young, R.J. Investigation of structure/property relationships in particulate composites through the use of Raman spectroscopy. *J. Raman Spectrosc.* **1999**, *30*, 929–938.
46. Ash, B. J.; Siegel, R. W.; Schadler, L. S. Mechanical Behavior of Alumia/Poly (methyl methacrylate) Nanocomposites. *Macromolecules* **2004**, *37*, 1358–1369.
47. Lipatov, I. S.; Lipatov, Y. S.; Knovel. Polymer Reinforcement. ChemTec Publishing, 1995.
48. Menczel, J.; Varga, J. Influence of nucleating agents on crystallization of polypropylene I. Talc as a nucleating agent. *J. Therm. Anal.* **1983**, *28*, 161–174.
49. Fujiyama, M.; Wakino, T. Structures and properties of injection moldings of crystallization nucleator-added polypropylenes I. Structure–property relationships. *J. Appl. Polym. Sci.* **1991**, *42*, 2739–2747.
50. Nielsen, L. E. Mechanical Properties of Polymers and Composites;. New York: Marcel Dekker, 1974.
51. Srivastava, I.; Yu, Z.; Koratkar, N. Viscoelastic characterization of graphene polymer composites, Advanced Science, Engineering and Medicine, in-press (2012).
52. Kandanur, S.; Rafiee, M.; Yavari, F.; Schrameyer, M.; Yu, Z.; Blanchet, T. and Koratkar, N. Suppression of wear in graphene polymer composites. *Carbon* **2012**, 50, 3178–3183.
53. Blanchet, T. A; Kennedy, F. E. Sliding wear mechanism of polytetrafluoroethylene (PTFE) and PTFE composites. *Wear* **1992**, *153*, 229–243.
54. Ricklin, S. Review of design parameters for filled PTFE bearing materials. *Lubrication Engineering* **1977**, *33*, 487–90.
55. Bahadur, S.; Tabor, D. The wear of filled polytetrafluoroethylene. *Wear* **1984**, *98*, 1–13.
56. Kandanaur, Sashi S. Ph.D. Thesis, Rensselaer Polytechnic Institute, Troy, NY, 2010.
57. Zandiatashbar, A.; Picu, C. R.; and Koratkar, N. Control of Epoxy Creep using Graphene. *Small* **2012**, 8, 1676–1682.
58. Zheng, D.; Tang, G.; Zhang, H.-B.; Yu, Z.-Z.; Yavari, F.; Koratkar, N.; Lim, S.-H.; Lee, M.-W. In-situ thermal reduction of graphene oxide for high electrical conductivity and low percolation threshold in polyamide 6 composites. *Compo. Sci. Tech.* **2011**, DOI:10.1016/j.compscitech.2011.11.014.
59. Qi, X.-Y.; Yan, D.; Jiang, Z.; Cao, Y.-K.; Yu, Z.-Z.; Yavari, F.; Koratkar, N. Enhanced elec-

trical conductivity in polystyrene nanocomposites at ultra-low graphene content. *ACS Appl. Mater. Inter.* **2011**, *3*, 3130–3133.

60. Balandin, A.; Ghosh, S.; Bao, W.; Calizo, I.; Teweldebrhan, D.; Miao F.; Lau, C. N. Superior thermal conductivity of single-layer graphene. *Nano Let.* **2008**, *8*, 902–907.
61. Cai, D.; Song, M. Recent advance in functionalized graphene/polymer nanocomposites. *J. Mater. Chem.* **2010,** *20*, 7906–7915.
62. Veca, L. M.; Meziani, M. J.; Wang, W.; Wang, X.; Lu, F.; Zhang, P.; Lin, Y.; Fee, R.; Connell, J. W.; Sun, Y. P. Carbon nanosheets for polymeric nanocomposites with high thermal conductivity. *Adv. Mater.* **2009**, *21*, 2088–2092.
63. Yu, A.; Ramesh, P.; Sun, X.; Bekyarova, E.; Itkis, M. E.; Haddon, R. C. Enhanced thermal conductivity in a hybrid graphite nanoplatelet—carbon nanotube filler for epoxy composites. *Adv. Mat.* **2008**, *20*, 4740–4744.
64. Gangulia, S.; Roya, A. K.; Anderson, D. P. Improved thermal conductivity for chemically functionalized exfoliated graphite/epoxy composites. *Carbon* **2008**, *46*, 806–817.
65. Yavari, F.; Fard, H. R.; Pashayi, K.; Rafiee, M. A.; Zamiri, A.; Yu, Z.-Z.; Ozisik, R.; Borca-Tasciuc, T.; Koratka, N. Enhanced thermal conductivity in a nanostructured phase change composite due to low concentration graphene additives. *J. Phy. Chem. C* **2011**, *115*, 8753–8758.
66. Han, M. Y.; Oezyilmaz, B.; Zhang, Y.; Kim, P. Energy bandgap engineering of graphene nanoribbons. *Phys. Rev. Lett.* **2007**, *98*, 206805.
67. Chen, Z.; Lin, Y.-M.; Rooks, M. J.; Avouris, P. Graphene nanoribbon electronics. *Phys. E* **2007**, *40*, 228–232.
68. Campos-Delgado, J.; Romo-Herrera, J. M.; Jia, X.; Cullen, D. A.; Muramatsu, H.; Kim, Y. A.; Hayashi, T.; Ren, Z.; Smith, D. J.; Okuno, Y.; et al. Bulk production of a new form of sp2 carbon: Crystalline graphene nanoribbons. *Nano Lett.* **2008**, *8*, 2773–2778.
69. Marcano, D. C.; Kosynkin, D. V.; Berlin, J. M.; Sinitskii, A.;Sun, Z.; Slesarev, A.; Alemany, L. B.; Lu, W.; Tour, J. M. Improved synthesis of graphene oxide. *ACS Nano* **2010**, *4*, 4806–4814.
70. Rafiee, M. A.; Lu, W.; Thomas, A. V.; Zandiatashbar, A.; Rafiee, J.; Tour, J. M.; Koratkar, N. Graphene nano-ribbon composites. *ACS Nano* **2010**, *4*, 7415–7420.
71. Suhr, J.; Zhang, W.; Ajayan, P.; Koratkar, N. Temperature activated interfacial friction damping in carbon nanotube polymer composites. *Nano Letters* **2006**, *6*, 219–223.

CHAPTER 3

Hybrid Graphene/Microfiber Composites

GRAPHENE fillers have demonstrated an ability to enhance the mechanical properties of a variety of polymer matrices, as illustrated in Chapter 2 of this book. Significant enhancements in the polymer's stiffness, strength, damping, fracture toughness, and fatigue resistance have been reported. Such nanocomposites can be considered as "two-phase" systems, with the graphene and the polymer chains comprising the two phases in the composite. By contrast, study of "three-phase" composites has not received the same level of attention. The three phases include graphene, the polymer matrix, and conventional microscale continuous fibers (e.g., Kevlar, glass or Graphite fibers). It should be noted that for high performance structural applications (e.g., in the aerospace or automotive industry), a two-phase nanocomposite without continuous microfibers to carry the load appears unlikely to provide sufficient mechanical stiffness or strength to be competitive. Therefore, it is important to investigate three-phase nanocomposites involving the combination of a graphene-modified matrix, together with conventional microfiber reinforcement, and compare its performance to the traditional microfiber-reinforced polymer composites that are used by industry.

Improved mechanical properties of such fiber-reinforced composites (FRCs) can have strong practical applications. FRCs, with their favorable strength-to-weight and stiffness-to-weight ratios, are replacing their metal counterparts in a variety of high performance structural applications [1–2]. However, the principal limitation of FRCs is their brittle failure and insufficient fatigue life, which results in deficiencies in terms of performance, cost, safety, and reliability of structural components [3–4]. Consequently, there is great interest in developing new concepts for fatigue-resistant FRC composite materials. Wind energy is one of the emerging industries where such new fatigue-resistant materials can have a high impact [5]. Wind turbine blades are typically composed of glass or carbon-fiber epoxy composites and are prone to fatigue failure due to large cyclic bending loads encountered by the blades during regular operation. Wind is the fastest growing energy technology (~$50 billion investment in 2008) on the globe [5] and enhancing the fatigue properties and the operating life of FRC materials used in wind turbine construction is, therefore, of great practical relevance.

The bulk of the material included in this chapter has been adapted from Reference [11], published by the author's group and his collaborators.

3.1. PROCESSING OF HIERARCHICAL GRAPHENE COMPOSITES

Bulk quantities of graphene platelets (GPLs) are necessary for the processing of such hierarchical FRC composites. A convenient method to mass-produce such GPLs is by the one-step thermal reduction and exfoliation of graphite oxide [6–10]. This method has been discussed in detail in Chapter 1. In this section, I will demonstrate how such graphene platelets can be combined with traditional E-glass fabric plies (bi-directional, twill weave, style 7725 from Fibreglast, USA) and a Bisphenol-A based epoxy matrix (Epoxy 2000 from Fibreglast, USA) to generate hybrid or hierarchical FRC nanocomposites. The first step in the fabrication of these composites is to disperse the as-produced GPLs in the thermosetting epoxy resin via ultra-sonication. This method has been discussed in detail in Chapter 2. Then, the GPL/epoxy blend is painted, layer-by-layer, on the glass-fibers. At least eight glass-fiber plies are typically used to construct the composite laminate. For smaller number of plies, the test results may not be representative of bulk laminates. After all the glass-fiber plies are laid up and wetted with the GPL-infused epoxy, a vacuum bag is placed over the system, and the sample is allowed to cure under vacuum for 24 hours at room temperature. During this process, excess epoxy is extracted out of the composite into an absorbent cloth inside the vacuum bag. Lastly, the composite is taken out of the vacuum bag and placed in an oven at 90°C for high temperature cure for four hours. By this method, the estimated volume fraction of the glass micro-fibers in the composite structure typically lies in the 0.6 to 0.8 range.

Figure 3.1(b) illustrates the structure [11] of the hybrid composite. Scanning electron microscopy (SEM) imaging showing the hierarchi-

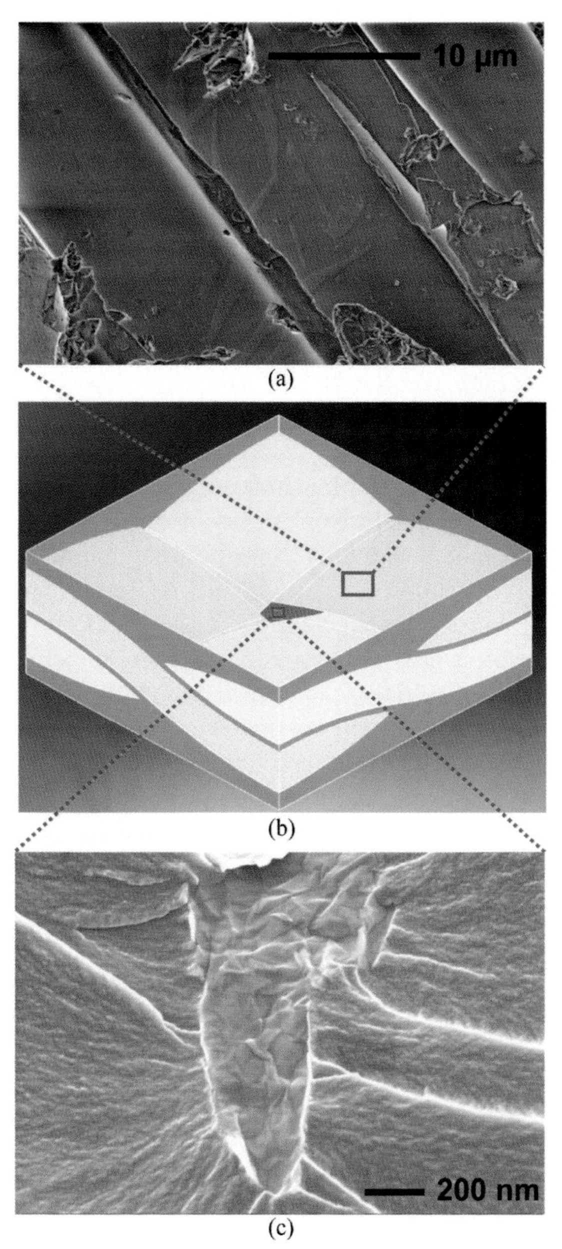

FIGURE 3.1. *Schematic representation of the unit cell that constitutes the 3-phase hierarchical composite. The three phases include interwoven E-glass strands laid up in the 0–90 direction, the epoxy matrix that serves as the binder, and GPLs dispersed into the epoxy matrix. Top scanning electron microscopy (SEM) image shows individual microscale glass fibers within the woven fabric strand. GPL additives interlinking the glass fibers through the epoxy matrix are also discernable. Lower SEM image shows a typical GPL/epoxy-matrix interface obtained from the matrix-rich region of the composite; there is no indication of interfacial debonding, which suggests a strong interface. (With permission from [11]).*

cal structure of the composite with the three main phases (i.e., the E-glass fibers, the epoxy resin, and the graphene platelets interlinking (or interlacing) the glass fibers) is seen in Figure 3.1(a). Figure 3.1(c) is a typical SEM image of the sample's fracture surface, indicating a graphene platelet embedded in the epoxy matrix. Although the sample was mechanically fractured, the integrity of the GPL/epoxy interface is maintained with no sign of debonding, suggesting a strong interface.

3.2. TESTING OF HIERARCHICAL GRAPHENE COMPOSITES

In a two-phase composite material, graphene is shown to enhance a broad range of mechanical properties of the host polymer matrix, such as Young's modulus, ultimate tensile strength, fracture toughness, fatigue crack propagation resistance, creep resistance, and wear resistance [12–18]. For a hierarchical three-phase composite with microfibers, the situation is not the same. Along the microfiber direction (i.e., laminate in-plane direction), the reinforcing ability of the distributed graphene platelets is negligible compared to the continuous microfibers that run along the length of the structure. In-plane laminate properties, such as Young's modulus and ultimate tensile strength, remain nearly identical despite the presence of graphene in the epoxy matrix. There is some improvement when one considers out of plane static mechanical properties, such as flexural bending stiffness and flexural strength. There is a ~20–30% increase in flexural strength (Figure 3.2), while the flexural modulus remained unchanged for the hierarchical and baseline composites. However, the situation changes dramatically when one considers the fatigue properties of the composite.

Fatigue test data on the hierarchical glass-fiber/epoxy/GPL laminates in a 3-point bend test configuration [11] can be generated as shown schematically in the inset of Figure 3.3. For the test data presented in this chapter, the cyclic loading tests were performed at a frequency of ~5 Hz and the stress ratio (R: minimum-to-maximum applied stress) was 0.1. Figure 3.3 shows the maximum flexural bending stress (S) vs. the number of cycles to failure (N) for the hierarchical composite for various GPL weight fractions up to a maximum of 0.2%. Note that 0.2% is the weight fraction of GPL in the epoxy resin. The weight fraction of GPLs in the entire laminate (including the E-glass microfibers) was estimated to be an order of magnitude lower (~0.02% for ~0.2% weight of GPLs in the epoxy). At each stress level, a minimum of three samples with the same GPL loading were tested to failure and the averaged results are represented as S-N curves (Figure 3.3). The results indicate a significant increase in the number of cycles to failure across the entire range of applied stresses. Increasing the GPL weight fraction from 0.05 to 0.2% had a strong beneficial impact on the fatigue life enhancement. At a stress level of ~500 MPa, the fatigue life of the composite with ~0.2% by weight of GPLs in the epoxy resin is enhanced by ~1200-fold as compared to the baseline glass-fiber/epoxy composite without the GPLs. At lower stress levels (< 400 MPa), about two orders of magnitude increase in fatigue life of the hierarchical composite relative to the baseline can be seen.

Figure 3.4 compares the performance of GPLs with SWNT and MWNT reinforcement at a constant nanofiller weight fraction of ~0.2%. The processing conditions used for SWNT and MWNT composites were identical to that of graphene. Depending on the applied stress, GPLs offer a 1–2 orders of magnitude increase in fatigue life as compared to MWNTs and SWNTs at the same weight fraction of additives. Figures 3.3–3.4 show flexural bending fatigue life results. Fatigue tests on hierarchical composites in the uniaxial tensile mode

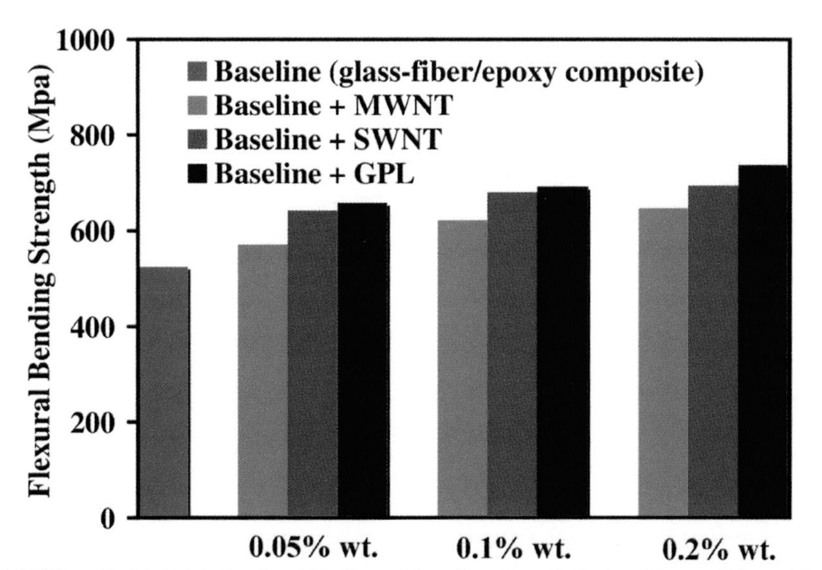

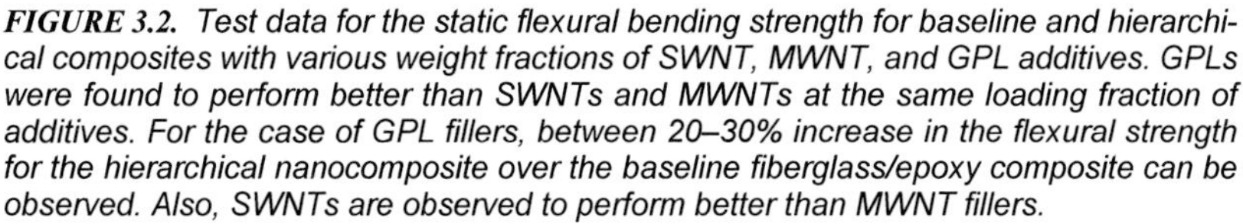
FIGURE 3.2. Test data for the static flexural bending strength for baseline and hierarchical composites with various weight fractions of SWNT, MWNT, and GPL additives. GPLs were found to perform better than SWNTs and MWNTs at the same loading fraction of additives. For the case of GPL fillers, between 20–30% increase in the flexural strength for the hierarchical nanocomposite over the baseline fiberglass/epoxy composite can be observed. Also, SWNTs are observed to perform better than MWNT fillers.

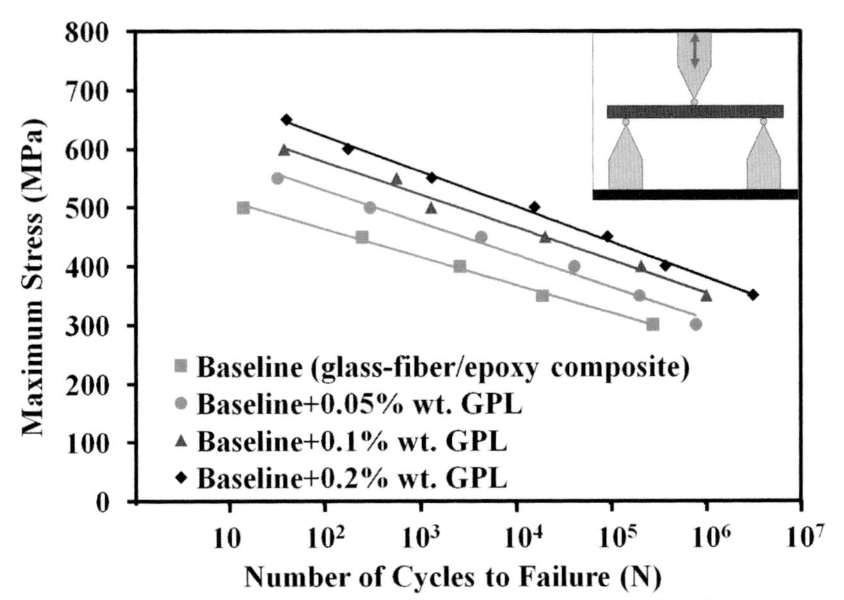

FIGURE 3.3. Fatigue test results in flexural bending mode. Maximum bending stress (S) vs. number of cycles to failure (N) for baseline glass-fiber/epoxy composites and glass-fiber/epoxy/GPL composites with various weight fractions of GPLs in the epoxy resin. The test is performed in the flexural bending mode as indicated in the inset schematic. (With permission from [11]).

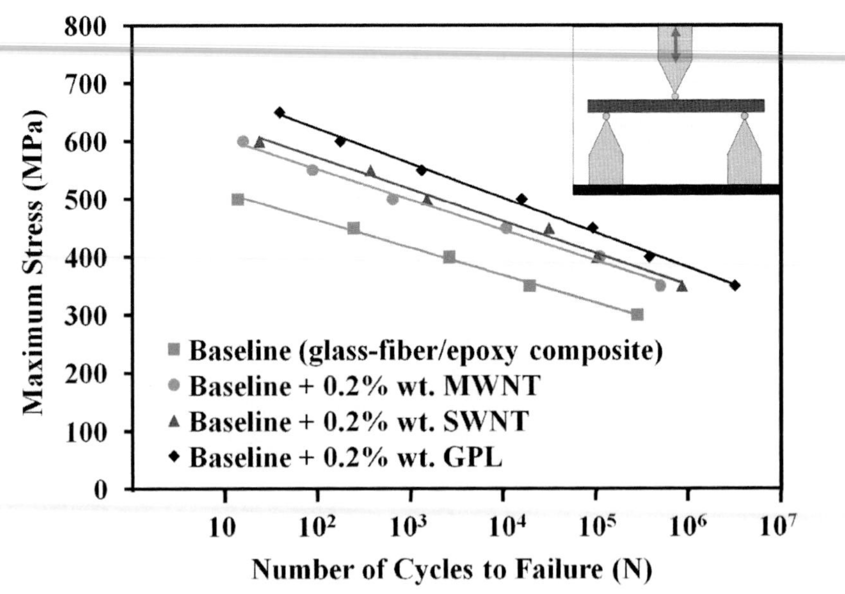

FIGURE 3.4. Fatigue test results in flexural bending mode. S-N curve comparing the fatigue response in flexural bending mode of GPLs with SWNT and MWNT additives at the same weight fraction of ~0.2% of the epoxy resin weight. (With permission from [11]).

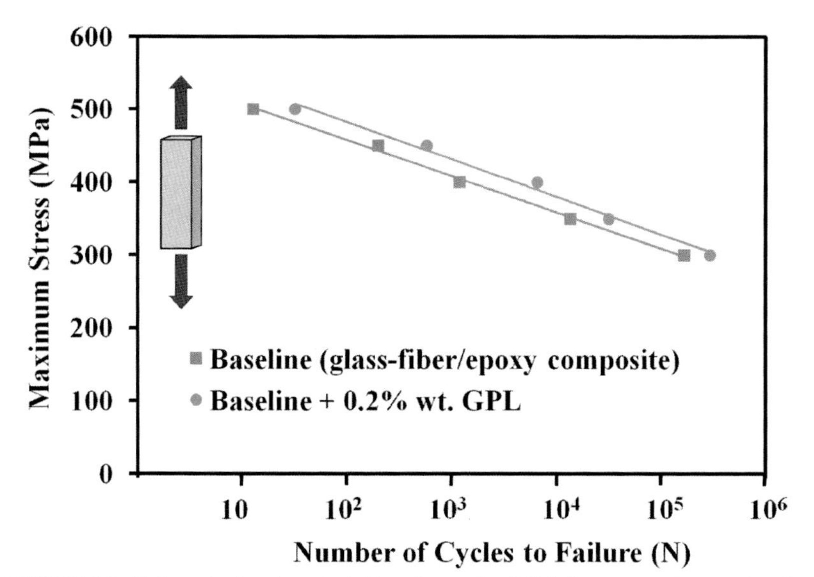

FIGURE 3.5. *Fatigue testing in uniaxial tensile mode. S-N fatigue curve in pure tension mode (i.e., no compressive stress) for baseline glass-fiber/epoxy and glass-fiber/epoxy/GPL composites. GPL weight fraction is ~0.2% of the epoxy resin weight. Inset schematic indicates that the fatigue tests were performed in the tensile loading mode. (With permission from [11]).*

(no compressive loading) are shown in Figure 3.5. For these cases, an enhancement in fatigue life is also observed, but the improvements are relatively modest (~3–5 times increase, depending on the stress level) compared to the flexural bending fatigue results shown previously.

The huge difference between the static and dynamic results for the hierarchical FRC composites, especially with respect to fatigue is quite striking. However, many studies have shown that nanofiller reinforcement influences dynamic mechanical properties such as fatigue far more than static properties [19–20], such as modulus and strength. For example, in Chapter 2, it was demonstrated that epoxy/graphene composites with ~0.1% weight of multi-layer graphene platelets enhanced the mode I fracture toughness of the baseline epoxy by ~53% (Figure 2.12), while the dynamic fatigue crack propagation rate (Figure 2.16) was suppressed by up to two orders of magnitude (~100-fold). In the hierarchical FRC system, if the graphene platelets are suppressing the propagation rate of inter-laminar fatigue cracks by ~100-fold, then it is conceivable that the total lifetime of the component (which involves the dynamic growth of such fatigue cracks and delaminations to critical dimensions) can also be enhanced by orders of magnitude.

To better understand the mechanism(s) responsible for the three orders of magnitude increase in fatigue life, ultrasonic C-scan imaging [21–22] of the samples prior to beginning the test, as well as during the fatigue test, is a useful tool. Figure 3.6(a) illustrates a typical ultrasound image of the longitudinal cross-section of the sample prior to fatigue loading. The image indicates a high degree of uniformity in the sample with absence of significant defects (voids) in the material. Figure 3.6(b) shows an ultrasound scan after 200,000 cycles of fatigue loading for the baseline glass-fiber/epoxy composite (no GPLs) at a maximum bending stress of ~300 MPa. The image shows a stark contrast between the tension side and the compression side of the sample. On the tension side, there is relatively less accumulation of damage to the sample, while on the compression side, a large increase in porosity is observed, suggesting delamination of the glass-fiber/epoxy-matrix interface. Interlaminar crack propagation [23–25] and the buckling of the glass-fibers under compressive load [26–27] typically cause such de-bonding. This results in large voids (i.e., damage) in the material, as seen in the ultra-

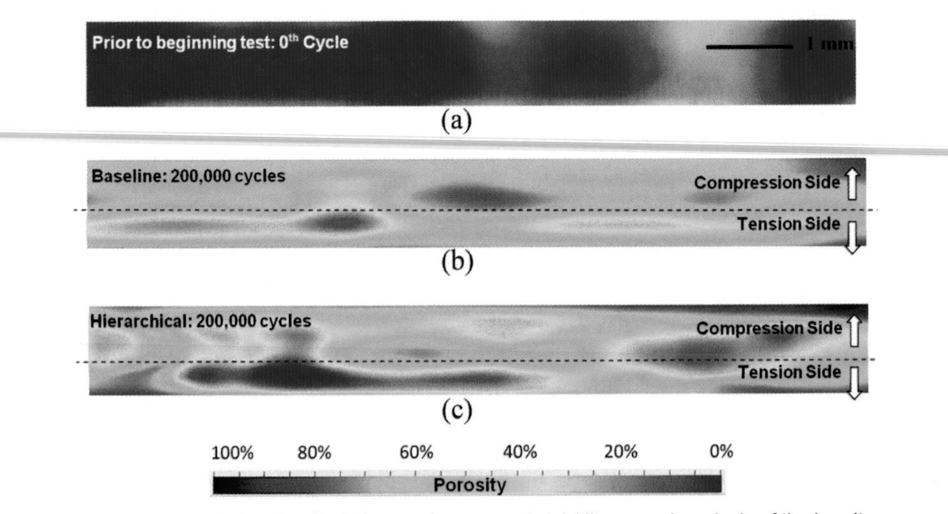

FIGURE 3.6. Mechanism for fatigue enhancement. (a) Ultrasound analysis of the longitudinal cross-section of a typical sample prior to fatigue loading. Red indicates 0% porosity, while blue indicates 100% porosity in the sample. (b) Ultrasound analysis of baseline glass-fiber/epoxy sample after 200,000 cycles of fatigue loading in bending mode at ~300 MPa peak bending stress. The tension (bottom) and compression (top) sides, as well as the neutral axis (dotted line) of the sample cross-section, are marked out for clarity. (c) Corresponding ultrasound results for glass-fiber/epoxy/GPL sample with ~0.2% weight of GPLs after 200,000 cycles of fatigue loading in bending mode at ~300 MPa peak bending stress. Scale bar indicated in Figure 3.6(a) also applies to Figures 3.6(b)-(c). (With permission from [11]).

sound images. The extent of damage on the compression side for the hierarchical glass-fiber/epoxy/GPL composite (also loaded for 200,000 cycles at ~300 MPa peak bending stress) is markedly lower than the baseline composite, as indicated in the ultrasound scan of Figure 3.6(c). The average porosity on the compression side after 200,000 cycles of fatigue loading for the hierarchical graphene composite is ~42%, compared to ~71% for the baseline composite. Moreover, in contrast to the baseline, there is no longer a large difference in "damage accumulation" between the tension side and the compression side of the sample during the flexural bending fatigue test. These results suggest that the main mechanism for prolonged fatigue life for the hierarchical composite appears to lie in the ability of the GPLs to suppress inter-laminar crack propagation and delamination/buckling of the glass-fiber/epoxy matrix interface under compressive stress. A similar phenomenon is also expected in the case of carbon nanotubes; however, this effect is far stronger for GPLs (see Figure 3.4), likely due to several reasons:

(1) GPLs have a rough and wrinkled surface topology, which can enable them to mechanically interlock with the glass-fibers and the epoxy matrix far more effectively than the atomistically smooth carbon nanotubes. Moreover, GPLs produced by thermal reduction of graphite oxide have residual hydroxyl and epoxide functional groups [28], which could interact covalently with the epoxy chains, thereby, further promoting interfacial adhesion.
(2) The specific surface area of GPL powder (> 700 m^2/g) is larger compared to nanotubes (in bulk powder form SWNTs and MWNTs show specific area in the range of 229–429 m^2/g [29]).
(3) The micron size dimensions, high aspect ratio, and two-dimensional sheet geometry of GPLs make them highly effective at deflecting cracks. GPLs have also been shown to be superior to SWNTs and MWNTs in terms of toughening the epoxy matrix (see Chapter 2).

Under pure tensile fatigue, the microfiber/matrix interface is less important because, unlike in compression, the microfibers are more effective [30–33] than the matrix in carrying tensile loads. Consequently, maintaining the integrity of the microfiber/matrix interface, while critical to boosting the fatigue life of the composite in bending/shear or direct compression, is relatively less important under pure tension (as observed in Figure 3.5). This analysis also suggests that the optimal placement of the GPL additives is right at the microfiber/matrix interface for maximum effectiveness. To test this hypothesis, we have explored an alternative method for fabrication of the hierarchical composites. In this method, GPLs are dispersed in acetone via ultra-sonication and then directly sprayed on to the glass-fibers. The regular epoxy (without any GPL additives) can then be applied, layer-by-layer, to the GPL-coated glass-fibers and cured. This method has two distinct advantages:

(1) There is a greater concentration of graphene sheets near the glass-fiber/epoxy interface, which increases the effectiveness of the graphene sheets.
(2) The epoxy without the nanofillers is less viscous and, therefore, easier to work with. If the graphene sheets are pre-mixed into the epoxy, the blend can become highly viscous, which makes it challenging to uniformly coat the microfibers with the nanocomposite resin.

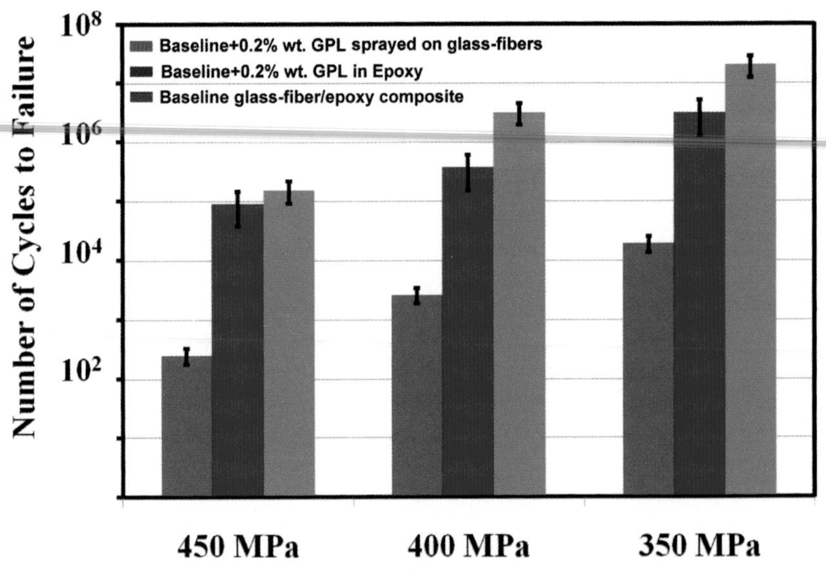

FIGURE 3.7. Flexural bending fatigue results comparing performance of 3-phase composites with GPLs dispersed into the bulk resin vs. the same amount of GPLs by weight directly spray-coated onto the glass microfibers. Data is shown at various peak bending stress levels. Directly spray-coating the GPLs at the fiber-matrix interface yields significant benefit over uniform dispersion of GPLs in the bulk epoxy resin. (With permission from [11]).

As expected, locating the GPL fillers closer to the glass-fiber/matrix interface results in significant enhancement in fatigue life (see Figure 3.7) over uniformly dispersing them in the bulk epoxy resin. For the case of 0.2% of GPLs (measured with respect to the epoxy resin weight required to cure the composite), but directly spray-coated on to the glass micro-fibers, the number of cycles to fatigue failure at a flexural bending stress of ~400 MPa is ~3000,000, which is ~8 times greater than when the GPLs are uniformly dispersed in the resin and ~1250 times greater than the fatigue life of the baseline glass-fiber/epoxy composite without any GPL reinforcement.

3.3. CONCLUSION

Hierarchical or hybrid fiber reinforced composites (e.g. glass-fiber/epoxy composites with various weight fractions of graphene platelets infiltrated into the epoxy resin as well as directly spray-coated on to the glass microfibers) can be manufactured. Remarkably only ~0.2% (with respect to the epoxy resin weight and ~0.02% with respect to the entire laminate weight) of graphene additives enhances the fatigue life of the composite in the flexural bending mode by up to 1200-fold. By contrast, under uniaxial tensile fatigue conditions, the graphene fillers resulted in ~3 to 5-fold increase in fatigue life. The fatigue life increase (in the flexural bending mode) with graphene additives was ~1–2 orders of magnitude superior to those obtained using carbon nanotubes. In-situ ultrasound analysis of the nanocomposite during the cyclic fatigue test suggests that the graphene network toughens the glass-fiber/epoxy-matrix interface and prevents the delamination/buckling of the glass microfibers under compressive stress.

The over three orders of magnitude enhancement in the flexural bending fatigue life of conventional microfiber-reinforced polymer composites achieved by the use of graphene platelets can translate into significant performance, cost, safety, and reliability benefits in a wide range of structural applications. The relatively low weight fractions of graphene additives (~0.2% of the epoxy resin weight and ~0.02% of the entire laminate weight) that are required, as well as the ability to directly spray-coat the graphene fillers on to the microfiber lamina, could make this concept particularly suitable for large-scale industrial applications in the aerospace, automobile, marine, and wind energy industries. However, to facilitate commercial applications, it is highly desirable to develop graphene pre-pregs with grapehene sheets pre-dispersed into a polymer matrix that is then coated on the individual glass/graphite/Kevlar lamina. Currently, such carbon nanotube pre-pregs are available commercially and several companies (e.g., Zyvex [34] in Columbus, Ohio, USA) are engaged in the development of graphene pre-pregs for hierarchical fiber-reinforced composite applications.

3.4. REFERENCES

1. Garg, A. C.; Mai, Y. Failure mechanisms in toughened epoxy resins—A review. *Compos. Sci. Technol.* **1988**, *31*, 179–223.
2. Argon, A. S.; Cohen, R. E. Toughenability of polymers. *Polym.* **2003**, *44*, 6013–6032.
3. Pearson, R. A.; Yee, A. F. Toughening mechanisms in elastomer-modified epoxies: Part 3. The effect of cross-link density. *J. Mater. Sci.* **1989**, *24*, 2571–2580.
4. Zhang, W.; Srivastava, I.; Zhu, Y.-F.; Picu, C. R.; Koratkar, N. Heterogeneity in epoxy nanocomposites initiates crazing: Significant improvements infatigue resistance and toughening. *Small* **2009**, *5*, 1403–1407.
5. Kong, C.; Bang, J.; Sugiyama, Y. Structural investigation of composite wind turbine blade considering various load cases and fatigue life. *Energy* **2005**, *30*, 2101–2114.
6. Ramanathan, T.; Abdala, A. A.; Stankovich, S.; Dikin, D. A.; Herrera-Alonso, M.; Piner, R. D.; Adamson, D. H.; Schniepp, H. C.; Chen, X.; Ruoff, R. S.; Nguyen, S. T.; Aksay, I. A.; Prud'Homme, R. K.; Brinson, L. C. Functionalized graphene sheets for polymer nanocomposites. *Nat. Nanotechnol.* **2008**, *3*, 327–331.
7. Schniepp, H. C.; Li, J.-L.; McAllister, M. J.; Sai, H.; Herrera-Alonso, M.; Adamson, D.H.; Prud'homme, R. K.; Car, R.; Saville, D. A.; Aksay, I. A. Functionalized single graphene sheets derived from splitting graphite oxide. *J. Phys. Chem. B* **2006**, *110*, 8535–8539.
8. McAllister, M. J.; Li, J-L.; Adamson, D. H.; Schniepp, H. C.; Abdala, A. A.; Liu, J.; Herrera-Alonso, M.; Milius, D. L.; Car, R.; Prud'homme, R. K.; Aksay, I. A. Single sheet functionalized graphene by oxidation and thermal expansion of graphite. *Chem. Mater.* **2007**, *19*, 4396–4404.
9. Rafiee, J.; Rafiee, M.; Yu, Z.-Z.; Koratkar, N. Superhydrophobic to superhydrophilic wetting control in graphene films. *Adv. Mater.* **2010**, *22*, 2151–2154.
10. Park, S.; Ruoff, R. S. Chemical methods for the production of graphenes. *Nat. Nanotech* **2009**, *4*, 217–224.
11. Yavari, F.; Rafiee, M. A.; Rafiee, J.; Yu, Z.; Koratkar, N. Dramatic increase in fatigue life in hierarchical graphene composites. *ACS Appl. Mat. & Inter.* **2010**, *2*, 2738–2743.
12. Rafiee, M. A.; Rafiee, J.; Wang, Z.; Song, H.; Yu, Z-Z.; Koratkar, N. Enhanced mechanical properties of nanocomposites at low graphene content. *ACS Nano* **2009**, *3*, 3884–3890.
13. Rafiee, M. A.; Rafiee, J.; Wang, Z.; Song, H.; Yu, Z-Z.; Koratkar, N. Fracture and fatigue in graphene nanocomposites. *Small* **2010**, *6*, 179–183.
14. Kandanur, S. S.; Rafiee, A. M.; Yavari, F.; Schrameyer, M.; Yu, Z.-Z.; Blanchet, T. A. ; Koratkar, N. Suppression of wear in graphene polymer composites. *Carbon* **2011**, *49*, 1258–1265.
15. Stankovich, S.; Dikin, D. A.; Dommett, G. H. B.; Kohlhaas, K. M.; Zimney, E. J.; Stach, E. A.; Piner, R. D.; Nguyen, S. T.; Ruoff, R. S. Graphene-based composite materials. *Nature* **2006**, *442*, 282–286.
16. Zhao, X.; Zhang, Q. H.; Chen, D. J.; Lu, P. Enhanced mechanical properties of graphene-based poly (vinyl alcohol) composites. *Macromolecules* **2010**, *43*, 2357–2363.
17. Fornes, T. D.; Paul, D. R.; Modeling properties of nylon 6/clay nanocomposites using composite theories. *Polymer* **2003**, *44*, 4993–5013.
18. Liang, J.J.; Huang, Y.; Zhang, L.; Wang, Y.; Ma, Y. F.; Guo, T. Y.; Chen, Y. S.; Molecular-

level dispersion of graphene into poly (vinyl alcohol) and effective reinforcement of their nanocomposites. *Adv. Funct. Mater.* **2009**, *19*, 2297–2302.

19. Wetzel, B.; Rosso, P.; Haupert, F.; Friedrich, K. Epoxy nanocomposites-fracture and toughening mechanisms. *Eng. Fract. Mech.* **2006**, *73*, 2375–2398.
20. Blackman, B.; Kinloch, A.J.; Sohn Lee, J.; Taylor, A.C.; Agarwal, R.; Schueneman, G.; Sprenger, S. The fracture and fatigue behaviour of nano-modified epoxy polymers. *J. Mater. Sci.* **2007**, *42*, 7049–7051.
21. Aymerich, F.; Meili, S. Ultrasonic evaluation of matrix damage in impacted composite laminates. *Compos. Part B-Eng.* **2000**, *31*, 1–6.
22. Kas, Y. O.; Kaynak, C. Ultrasonic (C-scan) and microscopic evaluation of resin transfer molded epoxy composite plates. *Polym. Test* **2005**, *24*, 114–120.
23. Talrega, R.; Ed. Damage Mechanics of Composite Materials. Elsevier: Amsterdam **1994**, *9*,139–241.
24. Kinloch, A. J.; Wang, Y.; Williams, J. G.; Yayla, P. The mixed-mode delamination of fibre composite materials. *Compos. Sci. Technol.* **1993**, *47*, 225–237.
25. Albertsen, H.; Ivens, J.; Peters, P.; Wevers, M.; Verpoest, I. Interlaminar fracture toughness of CFRP influenced by fibre surface treatment: Part 1. experimental results. *Compos. Sci. Technol.* **1995**, *54*, 133–145.
26. Ma, C.; Ji, L.J.; Zhang, R. P.; Zhu, Y. F.; Zhang W.; Koratkar, N. Alignment and dispersion of functionalized carbon nanotubes in polymer composites induced by an electric field. *Carbon* **2008**, *46*, 706–710.
27. Tadjbakhsh, I. G.; Wang, Y. M. Fiber Buckling in three-dimensional periodic-array composites. *Int. J. Solids Struct.* **1992**, *29*, 3169–3183.
28. Williams, G.; Seger, B.; Kamat, P. V. TiO_2-Graphene nanocomposites. UV-assisted photocatalytic reduction of graphene oxide. *ACS Nano* **2008**, *2*, 1487–1491.
29. Liu, C.; Chen, Y.; Wu, C-Z.; Xu, S-T.; Cheng, H-M. Hydrogen storage in carbon nanotubes revisited. *Carbon* **2010**, *48*, 452–455.
30. Wichmann, M. H. G.; Sumfleth, J.; Gojny, F. H.; Quaresimin, M.; Fiedler, B.; Schulte, K. Glass-fibre-reinforced composites with enhanced mechanical and electrical properties-benefits and limitations of a nanoparticle modified matrix. *Eng. Fract. Mech.* **2006**, *73*, 2346–2359.
31. Vlasveld, D. P. N.; Parlevliet, P. P.; Bersee, H. E. N.; Picken, S. J. Fibre–matrix adhesion in glass-fFibre reinforced polyamide-6 silicate nanocomposites. *Compos. Part A-Appl. Sci.* **2005** *36*, 1–11.
32. Tanaka, K.; Tanaka, H.; Stress-ratio effect on mode II propagation of interlaminar fatigue cracks in graphite/epoxy composites. ASTM STP 1997, 126–142.
33. Srivastava, I.; Proper, A.; Rafiee, M. A.; Koratkar, N. Three-phase textile nanocomposites: significant improvements in strength, toughness and ductility. *J. Nanosc. Nanotechno.* **2010**, *10*, 1025–1029.
34. Zyvex Technologies (www.zyvextech.com/)

CHAPTER 4

Graphene Ceramic and Graphene Metal-Matrix Composites

IN Chapters 2–3, I have discussed the possibilities associated with the incorporation of graphene additives in polymeric systems. Graphene could also benefit other categories of matrix materials, including ceramics and metals. In particular, structural ceramics are becoming increasingly relevant for high temperature applications. Among these, silicon nitride (Si_3N_4) is a high temperature-resistant ceramic (up to 1500°C), and is also considered the most reliable structural ceramic due to the formation of an interlocking microstructure of α-Si_3N_4 that is reinforced with long rod-like β-Si_3N_4 grains [1]. However, Si_3N_4 is still not widely used in many elevated temperature (> 1000°C) applications due to its overall low toughness properties in comparison to metals. Conventional ceramic matrix composites (CMCs) use one-dimensional fibers as the reinforcement phase, such as carbon fibers [2] or carbon nanotubes [3–4], and ceramic whiskers [5–6]. By contrast, processing of graphene-reinforced bulk ceramic composites is challenging due to the thermal stability limitations of graphene at high temperatures. Ceramics start to densify and sinter at temperatures > 1000°C, and Si_3N_4 is usually sintered at ~1800°C. This makes it challenging to incorporate graphene, which has low thermal stability at temperatures in excess of ~600°C [7]. To overcome this obstacle, one possibility is to use spark plasma sintering (SPS), which is a process that reduces the time at temperature from hours to minutes over conventional sintering methods, thus limiting thermally induced structural damage to the graphene platelets (GPLs) by avoiding long processing times at high temperatures. Similar methods can also be used to process GPL/metal-matrix composites. This chapter summarizes some efforts to develop and test

GPL/Si_3N_4 and GPL/Al composites [7–8] that have been reported in the literature. It also briefly describes GPL/Ni and GPL/Pt composites for solar cells. There is very limited data in this field, but based on preliminary findings, graphene ceramic and metal-matrix composites show outstanding potential for practical applications.

The bulk of the material included in this chapter has been adapted from References [7, 8, 44, 45] published by the author's group and his collaborators.

4.1. CERAMIC MATRIX COMPOSITES

In order to obtain uniform, densified microstructures of the nanocomposites, colloidal processing methods [9–11] are preferred to create homogenously dispersed particle systems in aqueous suspension. Bulk quantities of GPLs required for this can be produced by the rapid thermal expansion (> 2000°C min^{-1}) of graphite oxide [12–13] as discussed previously in Chapter 2. Such graphene platelets are typically composed on average of ~3–4 graphene sheets, with less than 2 nm total platelet thickness [14]. Figure 4.1(a)–(b) show typical scanning elec-

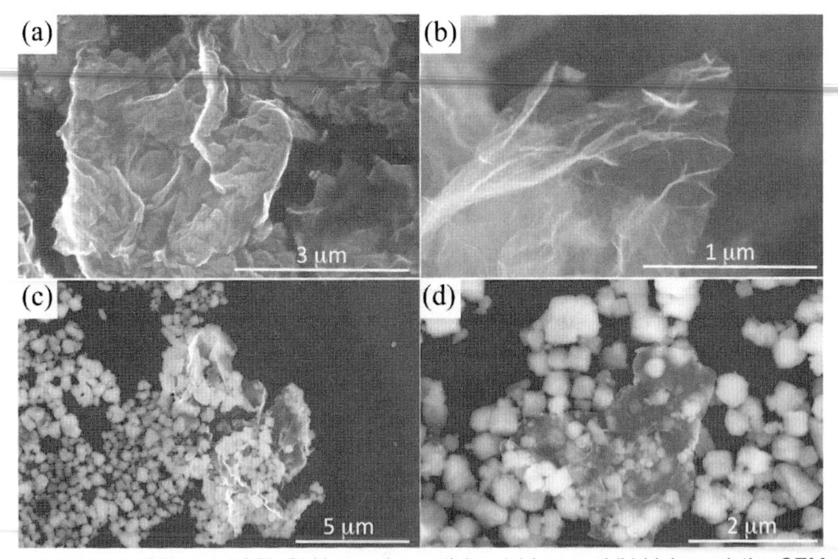

FIGURE 4.1. GPLs and GPL-Si_3N_4 powder particles. (a) Low and (b) high resolution SEM images of as-produced graphene platelets showing tightly packed platelets containing crumpled sheets of graphene. (c) Low and (d) high-resolution SEM images after colloidal processing, indicating partially exfoliated GPLs mixed with well-dispersed Si_3N_4 particles. The images clearly indicate GPLs decorated with Si_3N_4 particles; the Si_3N_4 particles are well dispersed throughout the surface area of the sheets. (With permission from [7]).

tron microscopy (SEM) images of the as-produced GPLs. The GPLs are agglomerated, as seen in Figure 4.1(a), before the colloidal processing step. In Figure 4.1(b), the higher resolution SEM image shows the wrinkled surface of a few partially separated graphene sheets. Cetyltrimethylammonium bromide (CTAB) can be used as a surfactant [15] to homogenize the GPLs and silicon nitride powder. One typically uses 1.0 wt. % CTAB to GPLs and 1.0 wt. % CTAB to Si_3N_4 powders in order to create a positive electrostatic repulsive force between the two phases of the composite materials and with each other. This electrostatic repulsion is developed due to the net charge from the positive head group on the surfactant molecules on the graphene platelets and the Si_3N_4 particles. Figure 4.1(c) and Figure 4.1(d) show SEM images of the GPLs and Si_3N_4 powder mixtures after aqueous colloidal processing. Figure 4.1(c) shows GPLs separated from each other among a uniform dispersion of Si_3N_4 particles. Figure 4.1(d) shows a higher resolution image of a GPL decorated with individual Si_3N_4 particles, which are also enveloped within the GPL.

Spark plasma sintering (SPS) is a relatively new high temperature powder consolidation method that has already been used to successfully create fully dense ceramics [16–18], nanoceramics [19–20], and ceramic nanocomposites [21–24] reinforced with carbon nanotubes. The advantages of using SPS to densify ceramics are: (1) rapid heating rates (up to 600°C/min); and (2) simultaneously applied pressure (60–120 MPa). SPS simultaneously applies pressure and quickly pulses electric current through a graphite die containing the ceramic powders that are to be densified. The pulsed current assists in densification upon applied pressure and relies on creep and related mechanisms for densification and not the conventional sintering methods that involve diffusion and mass transport of material across the grain boundaries during long periods of time at elevated temperatures. In this chapter, I will present data using SPS to densify GPL/Si_3N_4 nanocomposite powders with precise control of the matrix microstructure, and to limit the time at temperature to minimize the possibility of structural damage to the GPLs at high temperatures and pressures.

Table 4.1 shows the SPS heating rate, time at temperature, hold time, percent theoretical density (%TD), and final material composition obtained for monolithic Si_3N_4. Figure 4.2(a) shows the density plot for monolithic Si_3N_4 sintered from 1500 to 1700°C using 2 different hold times of 2 and 5 minutes at temperature. The 5-minute density plot shows that the density increases with increasing temperature from

TABLE 4.1. Physical Properties of Si_3N_4 Monoliths Densified Using SPS. (With permission from [7]).

Starting Material Composition	SPS Heating Rate (°C/min)	Sintering Temperature (°C)	Hold Time (min.)	Applied Load (MPa)	% Theoretical Density	Final Material Composition
α-Si_3N_4	100	1500	5	35	87.6	100% α-Si_3N_4
α-Si_3N_4	100	1575	5	35	98.4	100% α-Si_3N_4
α-Si_3N_4	100	1600	5	35	99.5	100% α-Si_3N_4
α-Si_3N_4	100	1700	5	35	99.0	83% α-Si_3N_4
α-Si_3N_4	100	1600	2	35	93.8	> 99% α-Si_3N_4
α-Si_3N_4	100	1625	2	35	97.0	> 99% α-Si_3N_4
α-Si_3N_4	100	1650	2	35	100.0	> 99% α-Si_3N_4

1500 to 1600°C (99.5% TD), and then remains relatively constant up to 1700°C (99.0% TD). The 2-minute density plot shows that the effect of temperature on densification is significantly greater when using a shorter hold time (keeping the heating rate constant) and results in 100% TD at ~1650°C. Figure 4.2(b) shows the shrinkage displacement curve and the heating profile curve that was measured during the SPS run for the high density monolithic part sintered at ~1650°C for 2 minutes. The displacement curve shows steady shrinkage displacement with increasing temperature up to ~1650°C, which represents ideal densification behavior. Figure 4.2(c) shows the X-Ray diffraction patterns for the monoliths sintered at various temperatures and hold times. Specifically, the XRD spectrum for the monolith sintered at 1650°C, for only 2 minutes, confirms that SPS can be used to tailor the matrix microstructure to ~100% α-Si_3N_4, while achieving high density. The accurate and reliable control of the matrix microstructure (α-Si_3N_4) is important because we want to evaluate the effect of GPLs as structural reinforcement within a uniform and homogenous equiax grain matrix microstructure. The XRD spectrum for the monoliths sintered at 1500, 1600, and 1700°C, for five minutes at temperature is also shown in Figure 4.2(c), with the percentage of each phase listed in Table 4.1. As expected, the phase content transitions from 100% α-Si_3N_4 to β-Si_3N_4 formation with increasing sintering temperature.

Figure 4.3(a) and Figure 4.3(b) are SEM images of two different fracture surfaces for the specimen of ~100% β-Si_3N_4 (> 99% T.D.). The grain size is estimated as ~500 nm and the grains are uniform throughout the fracture surfaces. Figure 4.3(c) and Figure 4.3(d) are SEM im-

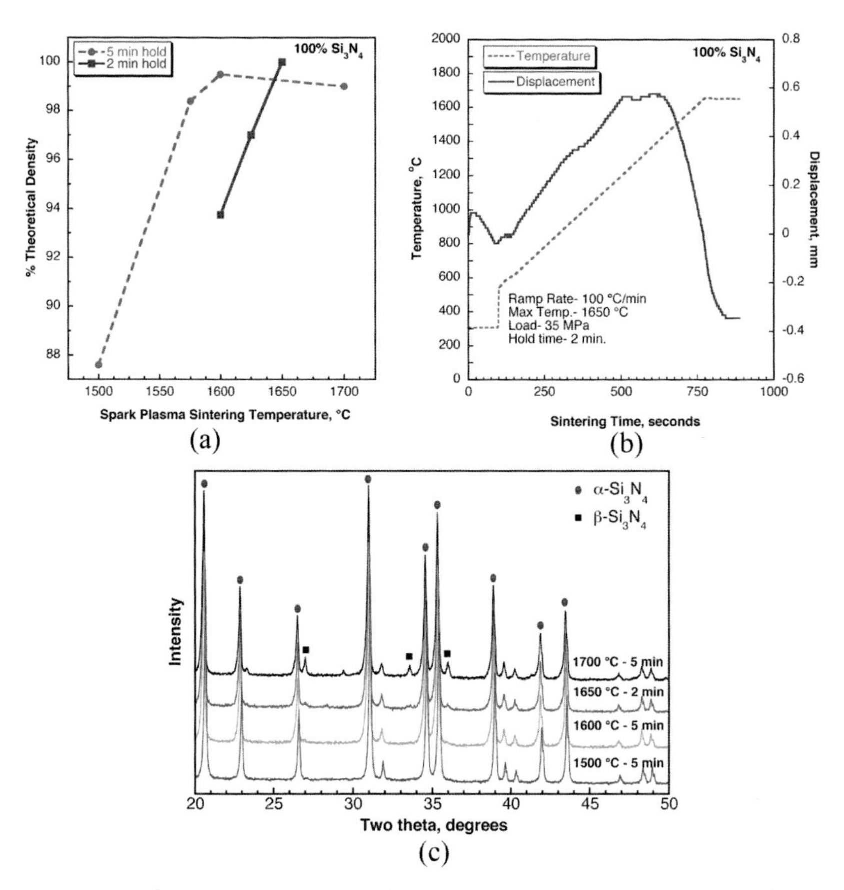

FIGURE 4.2. *Spark plasma sintering of Si_3N_4 matrix material. (a) The density of Si_3N_4 is plotted as a function of sintering temperature for two different times at temperature, which are 5-minute hold and 2-minute hold at temperature. 100% of theoretical density for the matrix material is obtained at ~1650°C, for 2-minute hold at temperature. (b) The densification profile for this high density part shows a steady increase in displacement with increasing temperature up to the final sintering temperature. (c) The X-ray diffraction phase analysis of the matrix after SPS shows that, for the highest density part (1650°C), we retain a phase composition that is approximately 100% α-Si_3N_4. (With permission from [7]).*

ages of two different fracture surfaces for the 1.0-vol% GPL-Si_3N_4 nanocomposite. Figure 4.3(c) is the lower resolution image showing the α-Si_3N_4 grain matrix microstructure. It also indicates homogeneous dispersion of the GPLs throughout the nanocomposite (as pointed out by the small white arrows directly labeled on the image). This specific image [Figure 4.3(c)] was taken at this area because of the interesting interactions between the large crack that runs through the bulk of

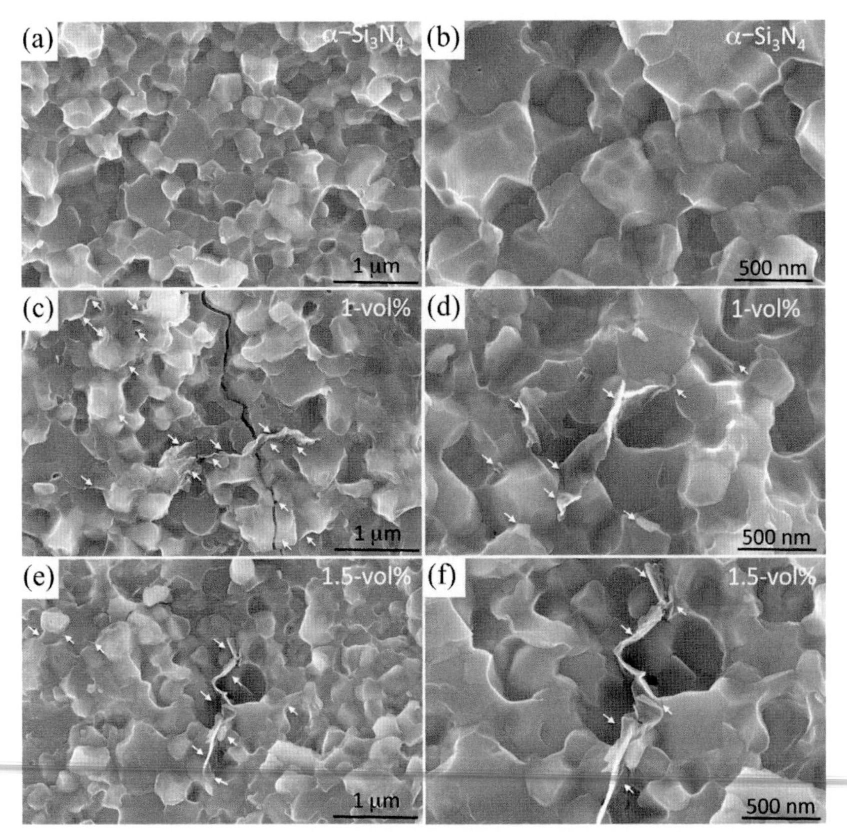

FIGURE 4.3. *SEM fracture surface images of Si_3N_4 and GPL nanocomposites. (a) Low and (b) higher magnification SEM images of the high-density, 100% α-Si_3N_4, monolithic ceramic matrix material showing an equiaxed and homogeneous grain microstructure. (c) Low and (d) higher magnification SEM images of the sintered and fractured, 1.0-vol% GPL-Si_3N_4 nanocomposite. The small white arrows illustrate the location of GPLs on the fracture surface image. The large crack running through image (c) indicates presence of GPLs along the grain boundary of the matrix material. Corresponding (e) Low and (f) high magnification SEM images of the 1.5-vol% GPL-Si_3N_4 nanocomposite. (With permission from [7]).*

the specimen and the GPL at the center of the image. First, we notice that the GPL is protruding out of the fracture surface and it is a large platelet that runs along the grain boundaries of the matrix. The long, continuous platelet of graphene does not appear to deflect the crack propagation path in-plane. However, the crack does not penetrate or puncture through the graphene platelet either. Therefore, it appears that the crack is arrested at the GPL and has to change directions (i.e., undergo out-of-plane deflection) to negotiate the GPL. Thus, it appears the GPLs (which are anchored at the grain boundaries) prevent cracks from changing their propagation paths in the conventional two-dimensional spaces and force such cracks to propagate in three-dimensional space. Such a fracture-resistance mechanism has hitherto not been reported in conventional ceramic matrix composite systems. Figure 4.3(d) is an SEM image of a fracture surface at a different location of the same nanocomposite (1.0-vol% GPL-Si_3N_4) and shows how the GPL (at the center of the image) is anchored securely within the grain boundaries of the matrix microstructure. The image also depicts smaller GPLs dispersed throughout the microstructure of the nanocomposite (pointed out by the small white arrows directly labeled onto the image).

Figure 4.3(e) and Figure 4.3(f) are SEM images of the same fracture surface for the 1.5-vol% GPL-Si_3N_4 nanocomposite. The bulk fracture surface is similar to the 1.0-vol% GPL-Si_3N_4 nanocomposite. At lower magnification [Figure 4.3(e)], one can clearly see the graphene sheets pulled out of the fracture surface. The higher resolution SEM image in Figure 4.3(f) illustrates a wall of graphene sheets that follows the grain boundaries of the matrix. Conventional fiber-reinforced ceramic toughening mechanisms, such as fiber pull-out, are commonly observed on fracture surfaces of bulk ceramic matrix composites. For the graphene system, one can also observe the pull-out of the graphene sheets that are tucked and wrapped around the matrix grains. The energy required to pull out a sheet is expected to be greater than that of a fiber due to "sheet-wrapping" around the matrix grain boundaries and the increased contact area with the matrix.

In order to qualitatively quantify the effect of graphene concentration on the toughness of the ceramic, microhardness testing can be used to induce radial cracking from the corners of the indentation. These cracks are then measured in length in order to calculate a toughness value using the Antis equation [25] [Equation (1)]. This equation uses the measured, hardness (H), applied load (P), modulus (E = 300 GPa, as measured previously for Si_3N_4), crack length (c_o), and a constant for Vickers-produced radial cracks in brittle ceramics (0.16) in order to calculate a toughness value.

$$K_C = 0.16\left(\frac{E}{H}\right)^{1/2}\left(\frac{P}{c_o^{3/2}}\right) \tag{1}$$

The Vickers hardness number (H) used to calculate the toughness values was measured using an applied load of 9.8 N in order to avoid forming radial cracks. An applied load of 98 N was used to create reproducible radial cracks that were used to measure crack values (c_o) used in Equation (1).

Figure 4.4(a) shows a representative microhardness indentation (inset image) of the 1.0-vol% GPL-Si_3N_4 nanocomposite. The area of the indent is approximately 150 μm^2 and was created using a 196 N applied load. Figures 4.4(a) and 4.4(b) are high-resolution images of the microhardness-induced radial cracks. Figure 4.4(a) shows crack deflection resulting in a branched crack structure. Probing within the cracks [Figure 4.4(a) and 4.4(b)], one can see direct evidence of "sheet pull-out" and graphene sheets that are bridging the cracks, which are directly labeled on the images. Figure 4.4(a) also shows two regions within the crack where it appears that the GPLs are necking down to a smaller cross-sectional area within the crack wake. Figure 4.4(c) shows the bulk fracture surface for the 1.0-vol% GPL-Si_3N_4 nanocomposite. The GPL at the center of the image is protruding out of the surface

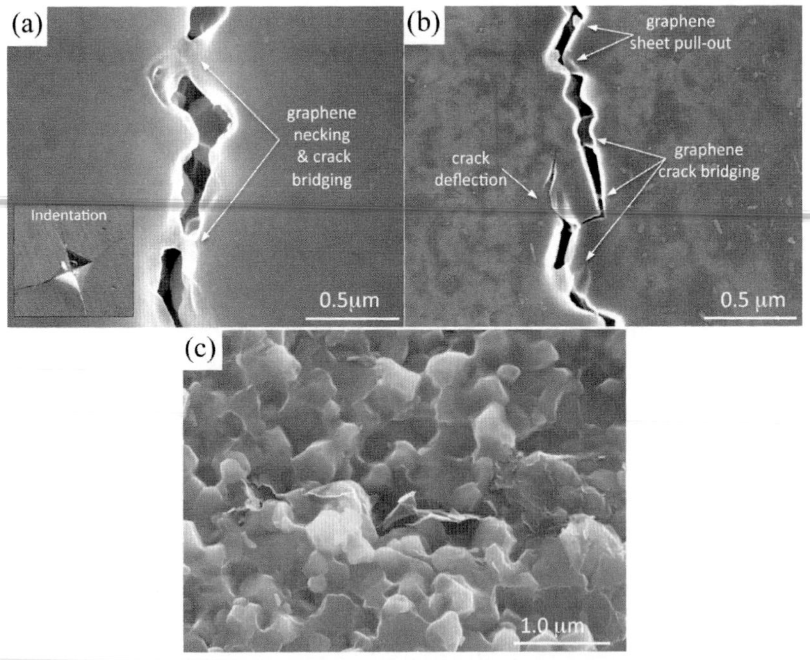

FIGURE 4.4. *Toughening mechanisms in GPL-Si_3N_4 nanocomposites. (a) Microhardness testing resulting in the creation of radial cracks stemming from the microhardness indent (inset image). Closer examination of the radial cracks reveals GPLs bridging the crack at several locations, two of which are shown in this high resolution SEM image. (b) Further examination of the radial cracks indicates they follow a tortuous crack propagation path. (c) Fracture surface of the bulk sample indicates the presence of three-dimensional toughening mechanisms for the GPL-Si_3N_4 nanocomposite. (With permission from [7]).*

and follows the contours formed by the matrix grain boundaries. This fracture surface again illustrates the ability of the GPL to block the in-plane propagation of the crack, thereby forcing it to climb over the wall of graphene sheets. Such a fracture surface is unexpected for a ceramic and suggests that the two-dimensional GPL promotes the deflection of cracks in three-dimensions.

Figure 4.5(a) is a plot of the calculated toughness values for the GPL-Si_3N_4 nanocomposites, shown as a function of GPL concentration from 0-, 0.02-, 0.5-, 1.0-, and 1.5-vol% GPL. The plot shows a systematic increase in toughness with increasing GPL concentration from ~2.8 to ~6.6 MPa-$m^{1/2}$. The increase in toughness over the monolith is as high as ~235% (i.e., three-fold increase in toughness for the 1.5-vol% GPL-Si_3N_4 nanocomposite over the monolith). The performance of GPLs is superior to single-walled carbon nanotube (SWNT) additives at the same filler volume fraction. For example, in [15], a fracture toughness of ~4.71 MPa$m^{1/2}$ for a 1.0-vol% SWNT-Si_3N_4 composite was reported, which is significantly lower than the values reported here (~5.8 MPa$m^{1/2}$) for GPLs. Fractography analysis (Figure 4.3–4.4) indicates the presence of a variety of toughening mechanisms for GPLs, including sheet wrapping, sheet pull-out, two- and three-dimensional crack deflection, and crack bridging. Table 4.2 summarizes the density, theoretical density, hardness, and toughness values for each nanocomposite (0.02-, 0.2-, 1.0-, and 1.5-vol% GPL-Si_3N_4) and the monolith that were all sintered at ~1650°C (for 2 minutes).

Raman study is essential to confirm that the sheet-like structures observed in Figures 4.3–4.4 are GPLs. Figure 4.5(b) shows a collection of

TABLE 4.2. Physical and Mechanical Properties of Graphene-Si_3N_4 Nanocomposites. (With permission from [7]).

Starting Material Composition	Density (g/cm^3)	% Theoretical Density	Hardness (GPa)	Toughness (MPa-m$^{1/2}$)
0.00 vol% Graphene + 100.00 vol% Si_3N_4	3.223	100.0	22.3 ± 0.84	2.8 ± 0.12
0.02 vol% Graphene + 99.98 vol% Si_3N_4	3.204	99.5	21.2 ± 0.34	2.7 ± 0.14
0.50 vol% Graphene + 99.50 vol% Si_3N_4	3.198	99.7	19.3 ± 0.69	5.21 ± 1.00
1.00 vol% Graphene + 99.00 vol% Si_3N_4	3.175	99.3	20.4 ± 0.37	5.8 ± 1.18
1.50 vol% Graphene + 98.50 vol% Si_3N_4	3.175	99.6	15.7 ± 0.61	6.6 ± 1.31

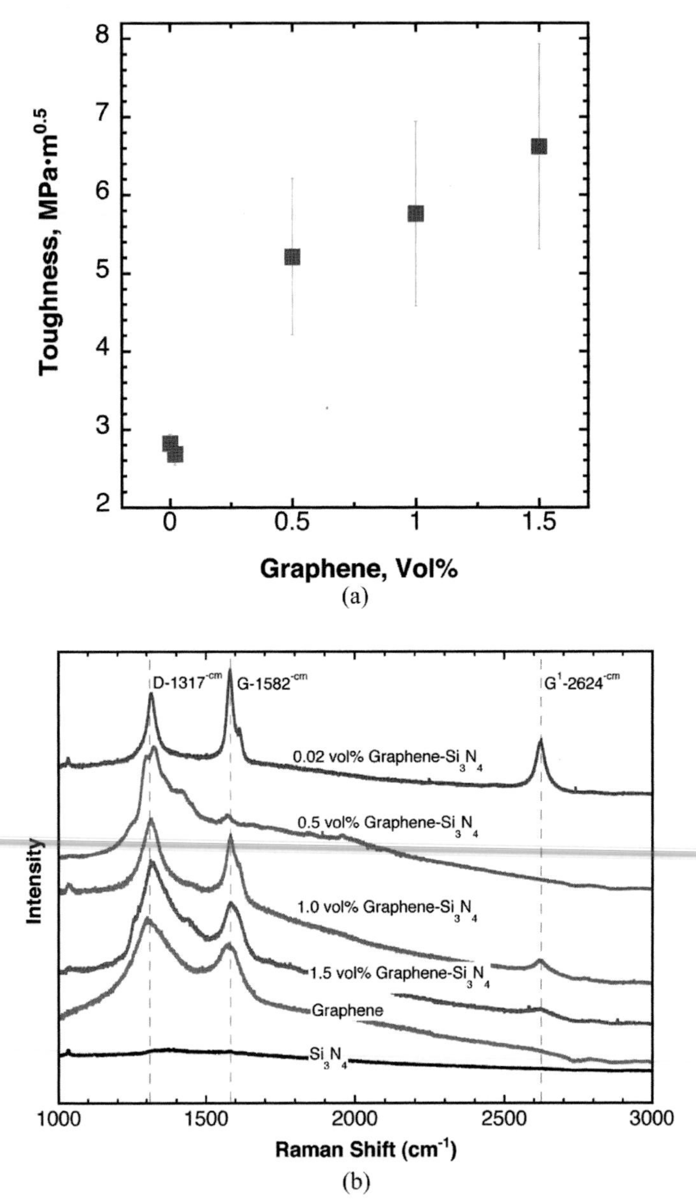

FIGURE 4.5. *Toughness characterization and Raman spectra for GPL-Si_3N_4 composites. (a) Toughness of the monolith systematically increases with increasing GPL vol. %, from 0–1.5 vol. % GPLs. The toughness of the monolith is enhanced by ~235% using ~1.5 vol. % GPL. (b) Raman spectroscopy was used to characterize the structure of the GPLs within the Si_3N_4 after SPS high temperature densification (1650°C for 2 minutes). Raman spectroscopy of the as-produced graphene shows the starting material as platelets and that the as-received starting powder of Si_3N_4 is not Raman active. At 0.02 vol. % GPLs, the SPS induces a transformation of the multilayer GPLs into few or bilayer graphene, which is detected by the appearance of a new peak (G′ band) at ~2624 cm^{-1}. At 0.5 vol. % GPLs, Raman signature of nanodiamonds is observed, while at 1.0- and 1.5 vol. % GPLs we detect a mixture of few-layered and multilayer graphene. (With permission from [7]).*

individual Raman spectra for as-produced graphene, Si_3N_4 (after sintering), and 0.02-, 0.5-, 1.0-, and 1.5-vol% GPL-Si_3N_4 nanocomposites (after sintering). The as-received graphene shows two clear peaks at ~1317 cm^{-1} (D-band) and ~1582 cm^{-1} (G-band). Note that the G′ peak at ~2624 cm^{-1} is absent, which is typical of multilayer sheets, or platelet configuration of graphene [26]. At ~0.02-vol% of graphene in the ceramic, the Raman signature after SPS indicates a new peak at ~2624 cm^{-1} for the G′ (or 2D band). This indicates the thinning of the multilayer graphene platelets into few-layer or possibly bilayer graphene. Therefore, the high temperatures and pressures associated with SPS can transform the structure of graphene from multilayers into few-layered graphene. However, as we increase the concentration from 0.02 to 1 and 1.5-vol% GPL, the intensity of the 2D peak diminishes, suggesting the graphene platelets are no longer being thinned as effectively at the higher loading fractions. Another interesting observation is that the Raman signature for the ~0.5-vol% GPL composite showed no peaks for the G and 2D bands. Instead, a new peak at ~1332 cm^{-1} appears corresponding to crystalline diamond. Clearly, further investigation is warranted to understand the effect of spark plasma sintering (i.e., pulsing direct current while under simultaneous pressure) on the thinning of GPLs and the possible conversion of GPLs to nanodiamonds.

To summarize, aqueous colloidal processing methods can be used to obtain uniform and homogenous dispersions of GPLs and Si_3N_4 ceramic particles prior to densification using SPS. After densification at ~1650°C, direct evidence of graphene in the nanocomposites can be found using Raman spectroscopy. At lower concentrations (0.02-vol% GPL), GPLs are converted into thinner bilayer or few-layered graphene sheets using SPS. An ~235% increase in toughness for the nanocomposite over the monolith is observed using only 1.5-vol% addition of graphene. Most interestingly, some very unexpected toughening mechanisms can be observed on the fracture surfaces of the nanocomposites. The GPLs appear to be anchored or wrapped underneath the matrix grains and result in the formation of a continuous wall of graphene along the grain boundaries that arrests and forces cracks to propagate in not just two- but in three-dimensions in order to continue to propagate through the material. Such crack deflection behavior in ceramics has hitherto not been reported. In addition to this, conventional toughening mechanisms such as filler pull-out and crack bridging are also observed in the case of graphene ceramic nanocomposites. Graphene nanofiller reinforcement could potentially be used to enhance toughness for a range of ceramic materials, enabling their widespread use in high performance structural applications.

4.2. METAL MATRIX COMPOSITES

A review of the literature indicates that, while graphene polymer composites [27–31] have been extensively researched, there has been little to no research in graphene metal matrix composites. This is likely a result of the greater difficulties in dispersion and fabrication, and the unknown interfacial chemical reactions in metal composites. This disparity in the amount of research given to polymer matrices as compared to metal matrices is seen in carbon nanotube (CNT) composites as well. Aluminum has been a common material to study in metal-carbon nanotube composites due to their diverse range of technical applications for lightweight alloys [32]. Researchers have seen mixed results, with some reporting little or no increase in mechanical strength while others have reported significant increases in strength. Many of these differences are a result of the quality of dispersion, fabrication method, and interfacial reactions that occur.

In this section, I will present data obtained by dispersion of graphene platelets in aluminum in order to observe the effects on mechanical strength. Valimet H-10 atomized pure aluminum powder with an average particle size of ~22 micron was used in this study. Bulk quantities of GPLs obtained by the one-step thermal exfoliation and reduction [12–13] of graphite oxide (see Chapter 2) were used to prepare the composites. Carbon vapor deposition (CVD) grown multi-walled carbon nanotubes (MWNTs) from NanoLab with ~15 nm diameter, ~15–20 micron length, and 95% purity were used to make nanotube samples to compare with the graphene performance.

To prepare the composites, the powders were blended, milled, pressed, and extruded. Aluminum-graphene composite powders were fabricated by initially blending the constituent precursory powders of Valimet Al and graphene. Blending was conducted using an acoustic mixer for ~5 minutes. This blend was then milled in a Zoz high energy attritor under an Argon atmosphere for one hour. Stearic acid was used as a process control agent to prevent agglomerations. In addition to creating a homogenous composite powder, the milling cycle also imparts some degree of grain refinement and breaks the nascent oxide layer off the aluminum, providing a clean metallurgical interface. This clean interface aided the consolidation process performed via instrumented hot isostatic pressing (I-HIP), uniquely equipped with a High Temperature Eddy Current Sensor (HiTECS) to monitor in real-time the densification of the composite powder. HIP conditions were tailored for each sample using the HiTECS, but processing was typically done at ~375°C for ~20 minutes. All samples were near 100% of the theoretical density as measured by the Archimedes method. Samples of pure aluminum, 0.1 wt. % graphene, and 1.0 wt. % MWNTs were made. Dispersion of graphene is more challenging compared to carbon nanotubes due to their greater interfacial contact area and, hence, a low weight fraction of graphene was intentionally selected. After hot isostatic pressing, the ~20 mm diameter billets were preheated to ~550°C for ~4 hours and then extruded on a 50-ton aluminum extrusion press. The extrusion ratio was 4:1, ram speed was ~12.5 mm/s, and extrusion pressure reached ~65 ksi. Pure aluminum, Al-graphene, and Al-MWNT samples were all prepared in the same manner.

Table 4.3 shows the hardness of the various samples after hot isostatic pressing and after extrusion. It is clear from the data that the aluminum reinforced with 1.0 wt. % MWNTs display the highest hardness among the materials tested. Pure aluminum showed an increase in hardness after extrusion to a value slightly below the nanotube-reinforced material. The increase in hardness is likely due to the formation of a more refined and compacted microstructure. The graphene composite showed high as-pressed hardness, but then exhibited a marked decrease in its hardness after extrusion. Figure 4.6 shows the tensile strengths of the extruded materials. The nanotube sample had the highest strength and the graphene sample showed the lowest tensile strength. The tensile strength of the nanotube composite was ~12% greater than the baseline, while the graphene composite showed ~18% lower tensile strength as compared to the baseline aluminum.

TABLE 4.3. Vickers Hardness Data for the Various Materials and Conditions. (With permission from [8]).

Material	Condition	Vickers Hardness
Pure Al	As-Pressed	83 ± 9
	As-Extruded	96 ± 7
Al-1.0 wt. % MWNT	As-Pressed	102 ± 4
	As-Extruded	102 ± 1
Al-0.1 wt. % Graphene	As-Pressed	99 ± 5
	As-Extruded	84 ± 5

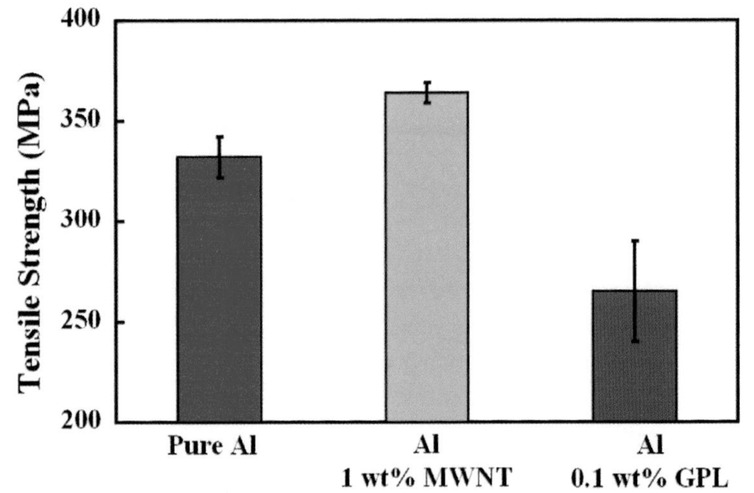

FIGURE 4.6. Ultimate tensile strengths of pure Al, Al-1 wt. % MWNTs, and Al-0.1 wt. % graphene. (With permission from [8]).

XRD scans for the Al, Al-0.1 wt. % graphene, and the 1.0 wt. % MWNTs are shown in Figure 4.7. All samples have major aluminum peaks at ~38.3° (111), ~44.6° (200), ~65.1° (220), ~78.2° (311), and ~82.3° (222). Despite only having ~0.1 wt. % graphene filler in the Al-graphene composite, strong peaks for aluminum carbide (Al_4C_3) are seen at ~31.2° and ~31.8°, ~55.0°, and ~72.5° two theta. Less intense peaks are seen in the other two samples. There may be some carbide that forms, even in the pure aluminum, from stearic acid/organics that

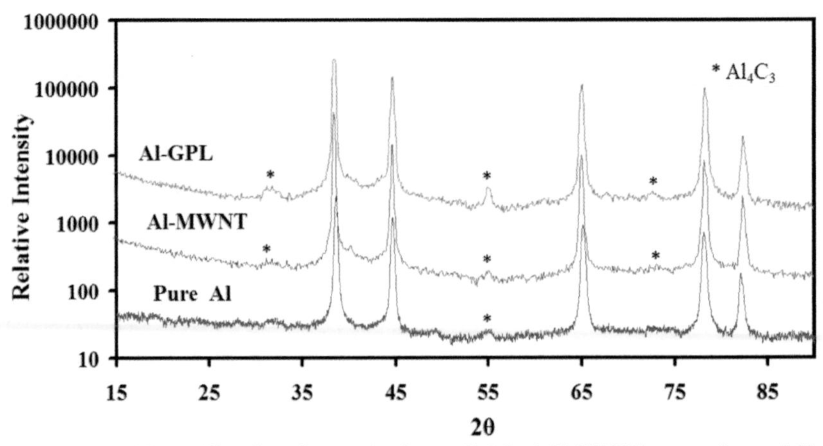

FIGURE 4.7. X-ray diffraction of pure aluminum, Al-1.0 wt. % MWNT composite, and Al-0.1 wt. % graphene composite after extrusion. (With permission from [8]).

may not have been fully removed [33] before HIP'ing. The graphene composite has the strongest peaks for Al_4C_3 among all the samples.

Aluminum carbide, Al_4C_3, which is the most energetically favorable stoichiometry of the aluminum carbides to form at the temperatures of interest in this study [34], will grow on the high surface free energy prismatic planes of carbon. This has been seen in conventional-sized carbon fibers and is deleterious to the strength of the composite [35–36]. The highly stable defect-free graphitic planes of the carbon nanotube or graphene do not react with aluminum to form aluminum carbide even at very high temperatures when the aluminum is liquid. Carbide formation will be promoted at defects in the graphitic planes (which exposes the prism planes), at tubes ends, and amorphous carbon coating at temperatures below the aluminum melting point [37]. Because of the complex conditions that lead to the growth or suppression of the aluminum carbide phase, some authors have reported the formation of Al_4C_3 [37–41], while others have not observed its formation [42–43].

The graphene platelets used in this research were produced by thermal reduction of graphite oxide. This processing results in graphene that has a wrinkled morphology and defects on the graphitic basal plane. This can be seen in the TEM micrograph in Figure 4.8(a). The abundance of defect sites is also confirmed by Raman spectroscopy. Raman analysis [Figure 4.8(b)] of the graphene powder indicated an intense D band and significant broadening of both the D and G bands, indicating a high degree of disorder. This high defect density is an artifact of the oxidation of graphite and the thermal shock technique that was employed to exfoliate graphite oxide to graphene platelets. The defects expose the prism planes of the graphene, which can become reaction sites with the aluminum. The abundant amount of prism planes at the graphene platelet edges could also become reaction sites. This could result in significant amounts of Al_4C_3 when compared to the total volume fraction of graphene since the graphene sheets are only on the order of a few atomic layers thick. Figure 4.9 shows the fracture surface of an Al-graphene tensile specimen. The microstructure shows the retained structure from powder consolidation and alignment during extrusion. Each grain, which is microns wide and nanometers thick, displays ductile failure in tension. If graphene platelets are adhered to the outside surface of the powders during milling and compaction and subsequently are transformed to aluminum carbide in large amounts, it could provide points of brittle weakness in-between the grains and cause the material to fail at lower strengths, as was observed.

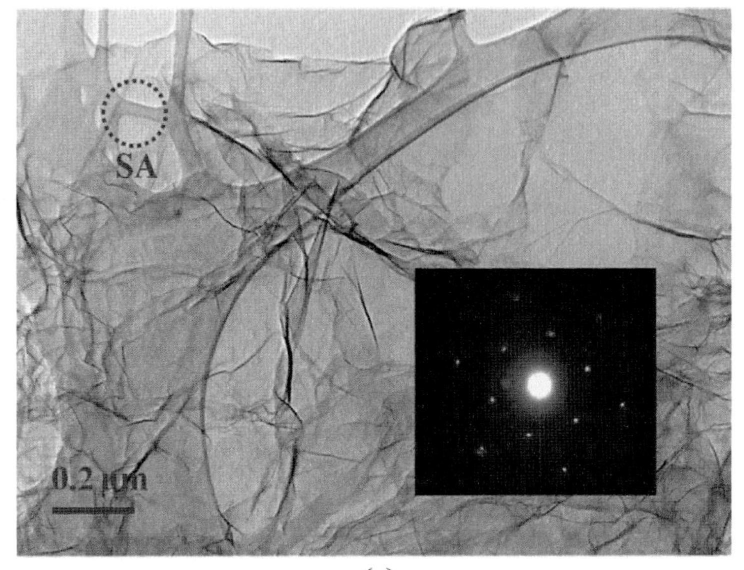

(a)

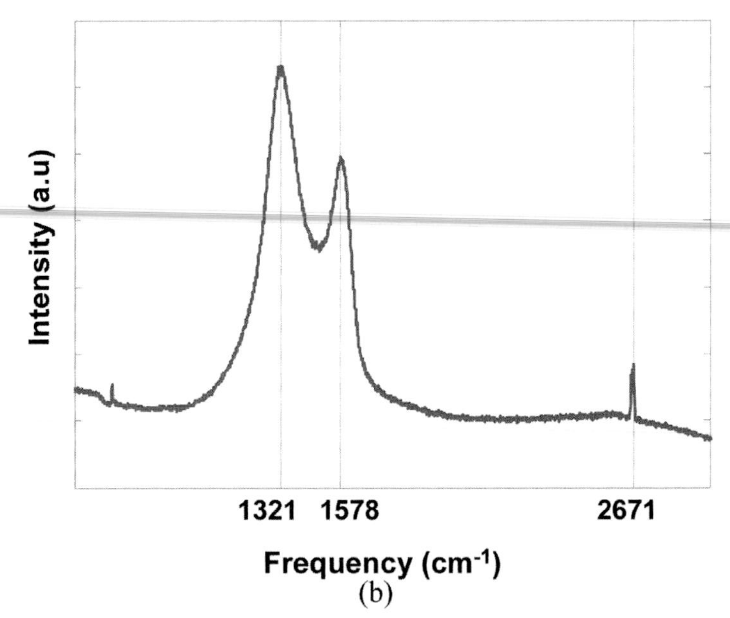

(b)

FIGURE 4.8. *(a) Transmission Electron Micrograph of graphene platelet showing the wrinkled morphology. Inset shows the selected area diffraction pattern (SADP) of the hexagonal graphene cell. (b) For Raman analysis, the graphene platelets were deposited on silicon wafers in powder form without using any solvent. Raman spectra of the samples were measured using a micro-spectrometer using an excitation wavelength of ~785 nm. The Raman G, D, and 2D band peaks are observed at ~1578 cm^{-1}, ~1321 cm^{-1}, and ~2671 cm^{-1}, respectively. Note that the D band peak is higher in intensity than the G band and both peaks are significantly broadened, which suggests a high density of defects in the graphene platelets. (With permission from [8]).*

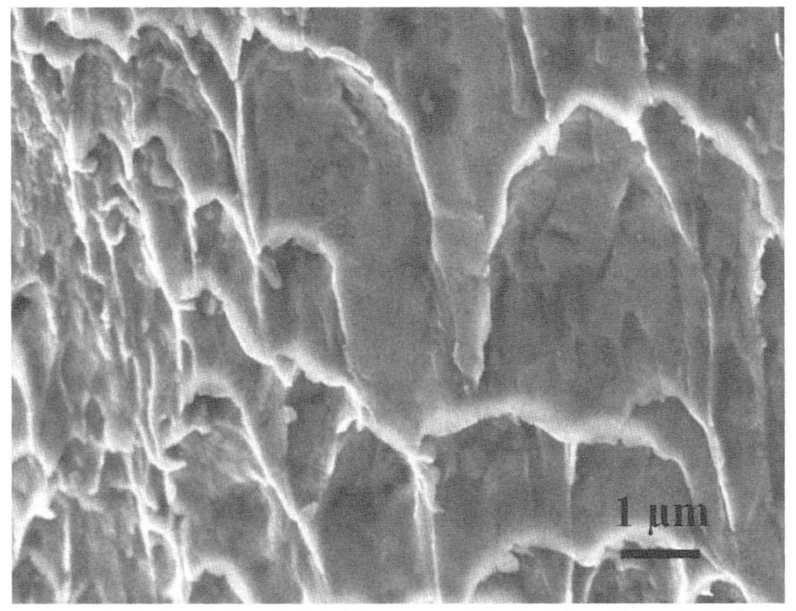

FIGURE 4.9. *Field-emission electron micrograph of the fracture surface of Al-0.1 wt. % graphene sample.*

To summarize, multi-walled carbon nanotubes can successfully increase the tensile strength of aluminum by up to ~12%. However, we find that graphene is prone to forming aluminum carbide during processing, which lowers the hardness and tensile strength of aluminum. The defective nature of graphene produced by thermal exfoliation/reduction of graphite oxide is likely responsible for the aluminum carbide formation. While defects in graphene have been shown to enhance interfacial binding and load transfer with polymer matrices, the same is not true for aluminum matrices. Therefore, it is challenging to process graphene-aluminum composites with good mechanical properties unless careful attention is given to the processing temperatures in order to avoid the formation of aluminum carbide. Graphene may still prove to be a promising reinforcement agent for metals, especially those that do not form a carbide, or ones in which very little carbide is formed, such as magnesium or copper. In this sense, the graphene could form reinforcing particles that could add strength to the composite in the same way that finely dispersed second phase precipitates do in precipitation hardened aluminum alloys.

Another interesting category of metal/GPL composites are GPL films coated with metal nanoparticles such as Pt or Ni. Pulsed laser deposition (PLD) techniques can be used to deposit Ni or Pt nanoparticles directly onto graphene films with great uniformity [44–45]. In this method, the third harmonic (355 nm) of Nd:YAG laser, with 10 Hz repetition rate and 5–6 ns pulse width, is used to ablate a Ni or Pt target. Laser light fluence is typically maintained at ~2.2 J.cm^{-2}. The Ni or Pt target is placed at a distance of ~48 mm from the substrate. Deposition time can be varied to control the number density of nanoparticles deposited on graphene. Substrate temperature is kept at ~450°C and pressure is ~10^{-6} mbar during the depositions. Figures 4.10–4.11 show the feasibility of depositing Pt and Ni nanoparticles [44–45] with great uniformity and control on GPLs. Pt nanoparticle-decorated graphene

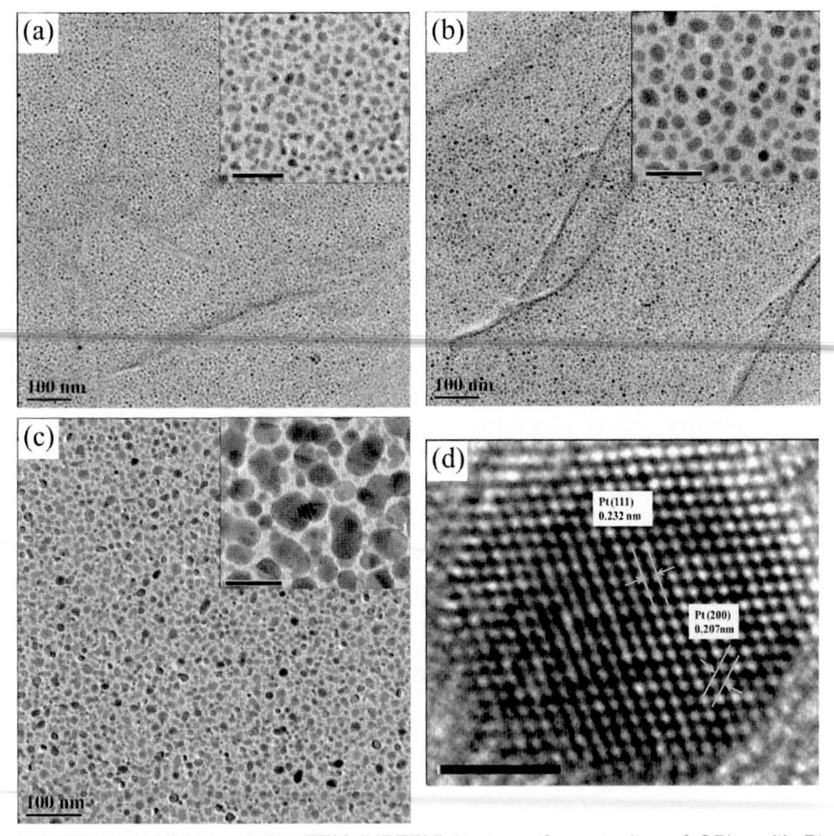

FIGURE 4.10. High resolution TEM (HRTEM) images of composites of GPLs with Pt nanoparticles having different Pt loadings (a) 15%, (b) 27%, and (c) 34% (scale bar is 100 nm). Inset shows corresponding high-resolution images (scale bar is 20 nm). (d) (111) and (200) lattice planes of a Pt nanoparticle (scale bar is 2 nm). (With permission from [44]).

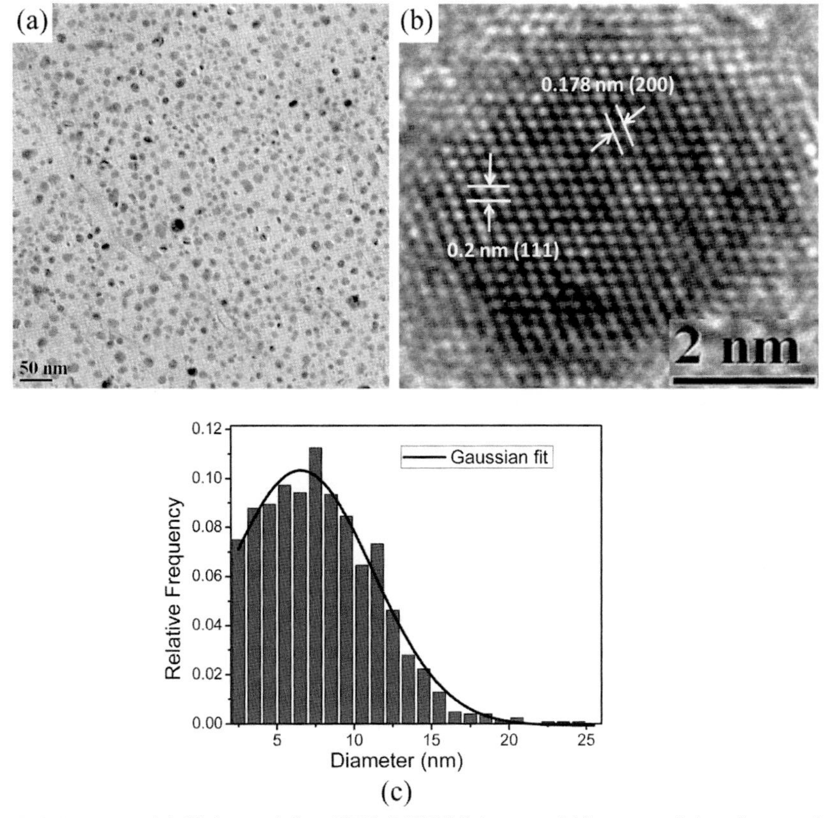

FIGURE 4.11. *(a) High resolution TEM (HRTEM) image of Ni nanoparticles dispersed uniformly over graphene sheets. (b) High resolution image showing lattice planes of Ni nanoparticles. (c) Average size distribution of Ni particles on the graphene film. (With permission from [45]).*

films have been tested as counter electrodes in dye sensitized solar cells and have achieved efficiencies as high as ~2.9%, which is significantly better than graphene alone (efficiency ~1%) or conventional Pt film-based electrodes, which show an efficiency of ~ 2%. Ni nanoparticle-decorated graphene films achieved efficiencies of ~2% in dye sensitized solar cells, which is comparable to conventional Pt film electrodes. It is expected that optimization of the nanoparticle size and surface coverage could potentially lead to further enhancements in the solar cell performance. Potential applications of the Pt nanoparticles supported on graphene also include conventional proton exchange membrane fuel cells, as well as emergent microbial fuel cells. Supporting Pt nanoparticles on graphene will facilitate electron transport between the Pt particles, while preventing the agglomeration of the Pt. This should significantly enhance electro-catalyst utilization and lower the mass loading of Pt in the electrode, which will improve the cost-effectiveness of the device.

4.3. REFERENCES

1. Riley, F. L. Silicon nitride and related materials. *J. Am. Ceram. Soc.* 2000, *83*, 245–265.
2. Hyuga, H.; Jones, M. I.; Hirao, K.; Yamauchi, Y. Fabrication and mechanical properties of Si_3N_4/Carbon fiber composites with aligned microstructure produced by a seeding and extrusion method. *J. Am. Ceram. Soc.* **2004**, *87*, 894–899.
3. Zhan, G. D.; Kuntz, J. D.; Wan, J.; Mukherjee, A. K. Single-wall carbon nanotubes as attractive toughening agents in alumina-based nanocomposites. *Nat. Mater.* **2002**, *2*, 38–42.
4. Zhang, T.; Kumari, L.; Du, G. H.; Li, W. Z.; Wang, Q. W.; Balani, K.; Agarwal, A. Mechanical properties of carbon nanotube–alumina nanocomposites synthesized by chemical vapor deposition and spark plasma sintering. *Compos. Part A* **2009**, *40*, 86–93.
5. Zhang, P.; Hu, P.; Zhang, X.; Han, J.; Meng, S. Processing and characterization of ZrB2-SiCW ultra-high temperature ceramics. *J. Alloy. Compd.* **2009**, *472*, 358–362.
6. Zhang, X.; Xu, L.; Du, S.; Han, W.; Han, J. Crack-healing behavior of zirconium diboride composite reinforced with silicon carbide whiskers. *Scripta Mater.* **2008**, *59*, 1222–1225.
7. Walker, L. S.; Marotto, V. R.; Rafiee, M. A.; Koratkar, N.; Corral, E. L.; Toughening in graphene ceramic composites. *ACS Nano* **2011**, *5*, 3182–3190.
8. Bartolucci, S. F.; Paras, J.; Rafiee, M. A.; Rafiee, J.; Lee, S.; Kapoor, D.; Koratkar, N. Graphene-aluminum nanocomposites. *Mater. Sci. Eng.* **2011**, *528*, 7933–7937.
9. Poyato, R.; Vasiliev, A. L.; Padture, N. P.; Tanaka, H. Aqueous colloidal processing of single-wall carbon nanotubes and their composites with ceramics. *Nanotechnology* **2006**, *17*, 1770–1777.
10. Sun, J.; Gao, L.; Li, W. Colloidal processing of carbon nanotube/alumina composites. *Chem. Mater.* **2002**, *14*, 5169–5172.
11. Blanch, A. J.; Lenehan, C. E.; Quinton, J. S. Optimizing Surfactant Concentrations for dispersion of single-walled carbon nanotubes in aqueous solution. *J. Phy. Chem. B* **2010**, *114*, 9805–9811.
12. Schniepp, H. C.; Li, J. L.; McAllister, M. J.; Sai, H.; Herrera-Alonso, M.; Adamson, D. H.; Prud'homme, R. K.; Car, R.; Saville, D. A.; Aksay, I. A. Functionalized single graphene sheets derived from splitting graphite oxide. *J Phy. Chem. B* **2006**, *110*, 8535–8539.
13. Rafiee J.; Rafiee M. A.; Yu Z.-Z.; Koratkar N; Superhydrophobic to superhydrophilic wetting control in graphene films. *Adv. Mater.* **2010**, *22*, 2151–2154.
14. Vadukumpully S.; Paul, J.; Valiyaveettil S.; Cationic surfactant mediated exfoliation of graphite into graphene flakes. *Carbon* **2009**, *47*, 3288–3294.
15. Corral, E. L.; Cesarano, J.; Shyam, A.; Lara-Curzio; E.; Bell, N.; Stuecker, J.; Perry, N.; Di Prima, M.; Munir, Z.; Garay, J.; Barrera, E V. Engineered Nanostructures for multifunctional single-walled carbon nanotube reinforced silicon nitride nanocomposites. *J. Am. Ceram. Soc.* **2008**, *91*, 3129–3137.
16. Munir, Z. A., Anselmi-Tamburini, U., Ohyanagi, M. The effect of electric field and pressure on the synthesis and consolidation of materials: A review of the spark plasma sintering method. *J. Mater. Sci.* **2006**, *41*, 763–777.
17. Garay J. E.; Current-Activated, Pressure-Assisted Densification of Materials. *Annu. Rev. Mater. Res.* **2010**, *40*, 445–468.
18. Hulberta, D. M.; Jianga, D.; Dudinaa, D. V.; Mukherjee, A. K. The synthesis and consolidation of hard materials by spark plasma sintering. *Int. J. Refrac. Met. H* **2009**, *27*, 367–375.
19. Vasylkiv, O.; Borodianskaa, H.; Sakkaa, Y. Nanoreactor engineering and SPS densification of multimetal oxide ceramic nanopowders. *J. Eur. Ceram. Soc.* **2009**, *28*, 919–927.

20. Casolco, S. R.; Xu, J.; Garay, J. E. Transparent/translucent polycrystalline nanostructured yttria stabilized zirconia with varying colors. *Scripta Mater.* **2008**, *58*, 516–519.

21. Zhan G. D.; Mukherjee A. K. Carbon nanotube reinforced alumina-based ceramics with novel mechanical, electrical, and thermal properties. *Int. J. Appl. Ceram. Technol.* **2004**, *1*, 161–171.

22. Dusza J.; Blugan G.; Morgiel J.; Kuebler J.; Inam F.; Peijs T.; Reece M. J.; Puchy V. Hot pressed and spark plasma sintered zirconia/carbon nanofiber composites. *J. Euro. Ceram. Soc.* **2009**, *29*, 3177–3184.

23. Estili M.; Takagi K.; Kawasaki A. Multiwalled carbon nanotubes as a unique agent to fabricate nanostructure-controlled functionally graded alumina ceramics. *Scripta Mater.* **2008**, *59*, 703–705.

24. Balazsi, C.; Shen, Z.; Konya, Z.; Kasztovszky, Z.; Weber, F.; Vertesy, Z.; Biro, L. P.; Kiricsi, I.; Arato, P. Processing of carbon nanotube reinforced silicon nitride composites by spark plasma sintering. *Compos. Sci. Technol.* **2005**, *65*, 727–733.

25. Anstis, G. R.; Chantikul, P.; Lawn, B. R.; Marshall, D. B. A critical evaluation of indentation techniques for measuring fracture toughness: I, direct crack measurements. *J. Am. Ceram. Soc.* **1981**, *64*, 533–538.

26. Dresselhaus, M. S.; Jorio A.; Hofmann M.; Dresselhaus G. Perspectives on carbon nanotubes and graphene Raman spectroscopy. *Nano Lett.* **2010**, *10*, 751–758.

27. Rafiee M. A.; Rafiee J.; Yu Z.; Koratkar N. Buckling resistant grapheme nanocomposites. *Appl. Phys. Lett.* **2009**, *95*, 223103.

28. Rafiee, M. A.; Rafiee, J.; Song, H.; Yu, Z.; Koratkar, N. Enhanced mechanical properties of nanocomposites at low graphene content. *ACS Nano* **2009**, *3*, 3884–3890

29. Rafiee, M. A.; Lu, W.; Thomas, A. V.; Zandiatashbar, A.; Raviee, J.; Tour, J.; Koratkar, N. Graphene nanoribbon composite. *ACS Nano* **2010**, *4*, 7415–7420.

30. Rafiee, M.; Rafiee, J.; Srivastava, I.; Wang, Z.; Song, H.; Yu, Z.-Z.; Koratkar N. Fracture and Fatigue in Graphene Nanocomposites. *Small* **2010**, *6*, 179–183.

31. Zhu, Y.; Murali, S.; Cai, W.; Li, X.; Suk, J.W.; Potts, J.R.; Ruoff, R. S.Graphene and Graphene Oxide: Synthesis, Properties and applications. *Adv. Mat.* **2010**, *22*, 3906–3924.

32. Kuzumaki, T. Processing of Carbon Nanetube Reinforced Aluminum Composite. *J. Mater. Res.* **1998**, *9*, 2445–2449.

33. Rubio-González, C.; Felix-Martinez, C.; Gomez-Rosas, G.; Ocaña, J.L.; Morales, M.; Porro J.A. Effect of laser shock processing on fatigue crack growth of duplex stainless steel. *Mater. Sci. Eng.* **2011**, *528*, 914–919.

34. Laha, T.; Kuchibhatla, S.; Seal, S.; Li, W.; Agarwal, A. Interfacial Phenomena in Thermally Sprayed Al-Based Nanocomposites Reinforced with Carbon Nanotubes. *Acta. Mater.* **2007**, *55*, 1059–1066.

35. Ishida, Y.; Ichinose, H.; Wang, J.; Suga, T. 46th Annu. Met. Electr. Micros. Soc. of America Proc/, edited by G. W. Bailey, San Francisco **1988**, 728.

36. Zhou, Y.; Yang, W.; Xia, Y.; Mallick, P.K. An experimental study on the tensile behavior of a unidirectional carbon fiber reinforced aluminum composite at different strain rates. *Mater. Sci. Eng.* **2003**, *A 362*, 112–117.

37. Ci, L.; Ryu, Z.; Jin-Phillipp, N.Y.; Ruhle, M. Investigation of the interfacial reaction between multi-walled carbon nanotubes and aluminum. *Acta. Mater.* **2006**, *54*, 5367–5375.

38. Deng, C.F.; Wang, D.Z.; Zhang, X.X.; Li, A.B. Processing and properties of carbon nanotubes reinforced aluminum composites. *Mater. Sci. Eng.* **2007**, *A 444*, 138–145.

39. Kwon, H.; Estili, M.; Takagi, K.; Miyazaki, T.; Kawasaki, A. Combination of hot extrusion and spark plasma sintering for producing carbon nanotube reinforced aluminum matrix composites. *Carbon* **2009**, *47*, 570–577.

40. Deng, C.F.; Zhang, X.X.; Wang, D.Z.; Ma, Y.X. Calorimetric study of carbon nanotubes and aluminum *Mater. Lett.* **2007**, *61*, 3221–3223.

41. Kwon, H.; Park, D.H.; Silvain, J.F.; Kawasaki, A. Investigation of carbon nanotube reinforced aluminum matrix composite materials. *Compos. Sci. Technol.* **2010**, *70*, 546–550.

42. Pérez Bustamante, R.; Estrada-Guel, I.; Antúnez-Flores, W.; Miki-Yoshida, M.; Ferreira, P.J.; Martínez-Sánchez, R. Novel Al-matrix nanocomposites reinforced with multi-walled carbon nanotubes. *Alloys Compds.* **2008**, *450*, 323–326.

43. George, R.; Kashyap, K.T.; Rahul, R.; Yamdagni, S. Strengthening in carbon nanotube/aluminum (CNT/Al) composites. *Scripta Mater.* **2005**, *53*, 1159–1163.

44. Bajpai, R.; Roy, S.; Kumar, P.; Bajpai, P.; Kulshrestha, N.; Rafiee, J.; Koratkar, N.; Misra. D. S. Graphene supported Platinum nanoparticle counter-electrode for enhanced performance of dye sensitized solar cells; *ACS Appl. Mater. Inter.* **2011**, *3*, 3884–3889.

45. Bajpai, R.; Roy, S.; Kumar, P.; Bajpai, P.; Kulshrestha, N.; Rafiee, J.; Koratkar, N.; Misra, D. S. Graphene supported Nickel nanoparticle as a viable replacement for Platinum in dye sensitized solar cells. *Nanoscale* **2012**, *4*, 926–930.

CHAPTER 5

Graphene Colloids and Coatings

ANOTHER important category of composites is nanofluids in which nanoscale particles form a stable colloidal dispersion in a fluid. Surface chemistry plays a critical role in the formation of a stable suspension in which the particles remain well dispersed in the liquid for long time duration. This chapter discusses the formation of colloids using graphene oxide, functionalized graphene oxide, and reduced graphene oxide, as well as surfactant stabilized graphene. It also discusses two potential applications of such colloids: (1) The spraying of nanofluids on surfaces to form graphene film coatings. Such coatings can dramatically influence the wettability of surfaces onto which they are coated. Controlling the wettability of solid surfaces is an important problem in surface engineering. (2) A very compelling application of graphene colloids–as a cutting fluid in micro-machining applications.

The bulk of the material included in this chapter has been adapted from References [4–5] published by the Han and Tascon Groups and References [3, 6, 35] published by the author's group and his collaborators.

5.1. GRAPHENE OXIDE COLLOIDS

Graphene oxide nanosheets are strongly hydrophilic due to the presence of a variety of oxygen-containing moieties, such as hydroxyl and epoxide functional groups [1–2]. Consequently, they form a very stable dispersion in water. The easiest way to generate such a colloid is to ultrasonicate graphite oxide flakes in water. Graphite oxide can be produced by the oxidation of graphite, as described previously in

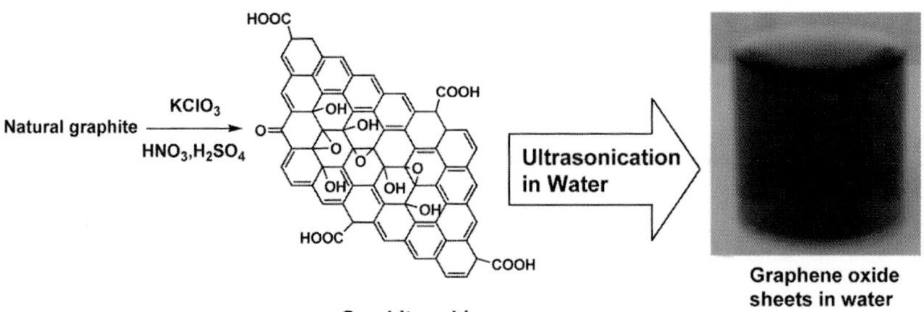

FIGURE 5.1. Oxidation of bulk graphite flakes (using modified Hummers Method) to yield graphite oxide. Vigorous ultrasonication in water results in exfoliation of the bulk graphite oxide into individual graphene oxide nanosheets. Water molecules are able to penetrate into the strongly hydrophilic graphite oxide layers and force open the structure. The stability of the dispersion is excellent and there is no indication of significant agglomeration for several months. A well-dispersed graphene oxide dispersion exhibits a characteristic pale yellowish-brown color. (Adapted with permission from [3]).

Chapter 1. After ~1 hour of ultrasonication using a high power probe sonicator, the graphene oxide sheets form a homogeneous colloidal suspension [3] with yellowish-brown color (Figure 5.1).

5.2. FUNCTIONALIZED GRAPHENE OXIDE COLLOIDS

The compatibility of graphene oxide with organic solvents can be controlled by functionalizing the graphene oxide surface. For instance, the covalent functionalization of graphene oxide with 2-amino-4, 6-di-dodecylamino-1, 3, 5-triazine (ADDT) causes a dramatic change in its

FIGURE 5.2. Solubility of (a) graphene oxide and (b) graphene oxide functionalized with ADDT in $CHCl_3/H_2O$ mixture (0.5mg/ml, 1.5 h later). (With permission from [4]).

wetting behavior [4]. The presence of two long alkyl chains in ADDT changes the polarity of graphene oxide, promoting its solubility in non-polar solvents. Figure 5.2 shows the solubility of graphene oxide and graphene oxide functionalized with ADDT in the mixture of $CHCl_3$/H_2O (v:v = 1/1) solvents. As $CHCl_3$ and H_2O are not miscible, $CHCl_3$ settled at the bottle of the glass vial. Due to its polar feature, the graphene oxide was completely dispersed in H_2O (the brown part of the right vial). On the contrary, when functionalized with ADDT, the graphene oxide was completely located and dispersed in $CHCl_3$ (the dark part of the left vial). The transformation of the super-hydrophilic graphene oxide nanosheets into a hydrophobic state by functionalization with ADDT is striking. This example illustrates vividly the powerful role played by surface chemistry in enabling stable colloidal suspensions of graphene and graphene oxide.

5.3. REDUCED GRAPHENE OXIDE COLLOIDS

Graphene oxide colloids can be transformed into graphene colloids by reduction of graphene oxide in solution. However, to avoid the re-stacking of the graphene sheets after reduction (due to the π–π interactions between sheets), covalent or noncovalent modification is needed by introducing extra stabilizers or surfactants into the reducing system. Alternatively, graphene colloids can be produced directly from graphene oxide suspensions without the use of a surfactant or stabilizer by a solvo-thermal method [4]. Briefly aqueous graphene oxide dispersions are centrifuged at ~11000 rpm for ~30 min. The supernatant liquid is then removed and the graphene oxide powder is re-dispersed in an organic solvent such as N,N-dimethylformamide (DMF). This process is typically repeated multiple times in order for proper solvent exchange to take place. To reduce the graphene oxide, the graphene oxide dispersion in DMF (0.5 mg mL^{-1}) is transferred into a stainless-steel autoclave (50 mL) and heated to a set temperature (~180°C) and maintained for ~12 h. After cooling to room temperature, the color of the dispersion changes from yellow-brown to deep black, indicating reduction of the graphene oxide nanosheets into graphene (Figure 5.3). The high temperature in the autoclave, coupled with the weak reducing effect of the DMF, is responsible for the reduction of the graphene oxide sheets. Such solvo-thermally reduced graphene (STG) dispersion in DMF is very stable and does not aggregate after more than one year [4], with concentrations of up to ~0.3 mg mL^{-1}.

To understand why graphene disperses well in DMF, one has to consider the solubility parameters of the solvent used. Hansen solubility parameters are typically used for such analysis. There are three Hansen solubility parameters [4]–dispersion cohesion parameter (δ_d), polarity cohesion parameter (δ_p), and the hydrogen bonding cohesion parameter (δ_H). It has been reported that graphene can be dissolved in solvents that have a value of ($\delta_p + \delta_h$) in the range of 13–29. DMF has a value of $\delta_p + \delta_h = 25$, which lies in this range.

The solvent exchange method can also be used to re-disperse the STG in other solvent systems. For this, ultra-pure water (with the same volume as the STG dispersion) is added into the STG dispersion in DMF. STG, being hydrophobic, precipitates out and can be collected by filtration [4]. The extracted graphene is typically dried in vacuum at ~60°C overnight. The solid STG can then be re-dispersed in a variety of solvents by sonication (160 W power) for ~1 h. As shown in Figure 5.4, in addition to DMF, graphene sheets shows good dispersibility in N-methyl-pyrrolidone (NMP), pyridine, N,N-dimethylacetamide (DMAC), and dimethyl sulfoxide (DMSO), as well as acetonitrile, with concentrations as high as ~1 mg mL^{-1}. This result is consistent with

FIGURE 5.3. *Photographs of (a) graphene oxide nanosheets in DMF (yellow-brown, 0.5 mg mL^{-1}), (b) STG in DMF (deep black, 0.3 mg mL^{-1}), and (c) STG in DMF (pale black, 0.01 mg mL^{-1}). (Adapted with permission from [4]).*

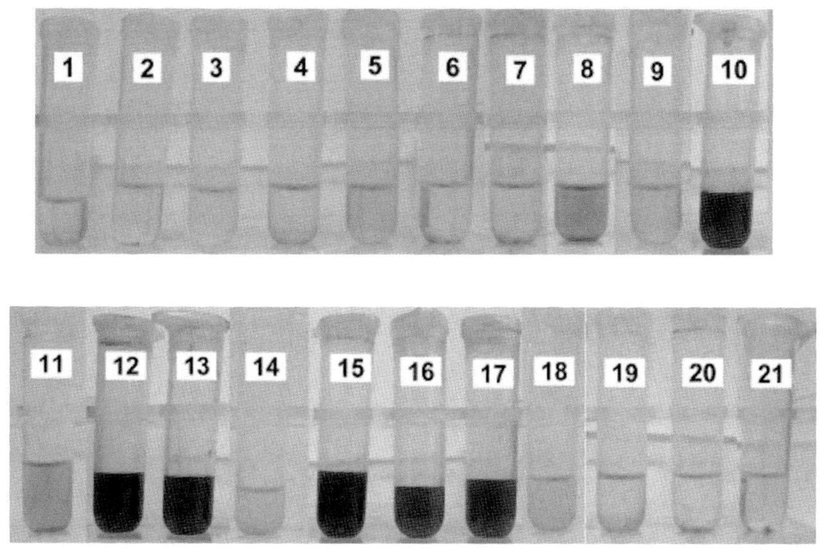

FIGURE 5.4. Dispersion test experiment of STG: Solvents 1-21 represent (1) hexane, (2) TEA, (3) toluene, (4) diethyl ether, (5) DEA, (6) DCM, (7) chloroform, (8) ethyl acetate, (9) THF, (10) pyridine, (11) acetone, (12) NMP, (13) DMAC, (14) 2-isopropanol, (15) acetonitrile, (16) DMF, (17) DMSO, (18) ethanol, (19) ethylene glycol, (20) water, and (21) methanol. (Adapted with permission from Ref. [4]).

the Hansen solubility parameters analysis presented earlier, since the $\delta_p + \delta_h$ value for all these solvents lies in the 12.7 to 28.7 range. STG is also dispersible [4] to a lesser degree in ethyl acetate, tetra-hydrofuran (THF), acetone, 2-isopropanol, ethanol, and ethylene glycol, and is not dispersible in hexane, triethylamide (TEA), toluene, diethyl ether, diethylamide (DEA), dichloromethane (DCM), chloroform, water, and methanol.

5.4. GRAPHENE COLLOIDS STABILIZED BY SURFACTANTS

Guardia and co-workers [5] have shown that it is possible to directly exfoliate bulk graphite into graphene sheets by ultrasonicating them in a suitable surfactant solution. They investigated a variety of surfactants, both ionic as well as non-ionic (Table 5.1). The initial concentration of graphite in the surfactant solutions was ~100 mg mL^{-1}. Ultrasonication was performed by means of an ultrasound bath cleaner (JP Selecta Ultrasons system, 40 kHz) for ~2 h. Following sonication, the dispersions were centrifuged at 5000 g for 5 min to sediment unexfoliated particles

TABLE 5.1. List of Ionic and Non-ionic Surfactants and Their Acronyms. (Adapted with permission from [5]).

Surfactant Name	Acronym
Non-Ionic	
Pluronic® P-123	P-123
Tween 80	
Brij 700	
Gum arabic from acacia tree	
Triton X-100	
Tween 85	
Brij 30	
Polyvinylpyrrolidone	PVP
n-Dodecyl β-D-maltoside	DBDM
Ionic	
Poly(sodium 4-styrenesulfonate)	PSS
3-[(3-Cholamidopropyl)dimethyl ammonio]-1-propanesulfonate	CHAPS
Sodium deoxycholate	DOC
Sodium dodecylbenzene-sulfonate	SDBS
1-Pyrenebutyric acid	PBA
Sodium dodecyl sulphate	SDS
Sodium taurodeoxycholate hydrate	TDOC
Hexadecyltrimethylammonium bromide	HTAB

or thick flakes of graphite, and the top ~65% of supernatant, which was the final graphene dispersion, was collected.

Figure 5.5(a)–(f) show vials containing aqueous graphene dispersions stabilized by a few representative surfactants. The various surfactants show significant differences in their ability to disperse the graphene sheets. For example, the detergent TDOC yields a faint (transparent) dispersion [Figure 5.5(b)], suggesting very low grapheme concentrations. By contrast, the non-ionic surfactants Brij 700 and P-123 exhibit completely black and opaque suspensions [Figure 5.5(e)–(f)], indicating their ability to disperse large amounts of exfoliated graphene sheets [5]. The dispersions were observed to be stable for at least several weeks, showing little, if any, signs of precipitated material during such period of time. The graphitic nature of the suspended material in the suspension can be confirmed using UV-vis absorption spectroscopy [Figure 5.5(g) shows the typical spectra for a dispersion stabilized by P-123]. The absorption peak is located at ~269 nm, and can be attributed to the $\pi \rightarrow \pi^*$ transitions of aromatic C-C bonds.

To obtain quantitative estimates of the concentration of graphene sheets in the suspensions, Guardia and co-workers [5] measured the absorbance of the suspensions at a specific wavelength (660 nm) and estimated the corresponding concentrations [Figure 5.5(h)] using the known extinction coefficient for graphene dispersions in surfactant/water solutions ($\alpha = 1390$ mL mg^{-1} m^{-1}). The most interesting observation is that non-ionic surfactants are far more effective in the stabilization of graphene compared to ionic ones. The concentrations achieved for ionic

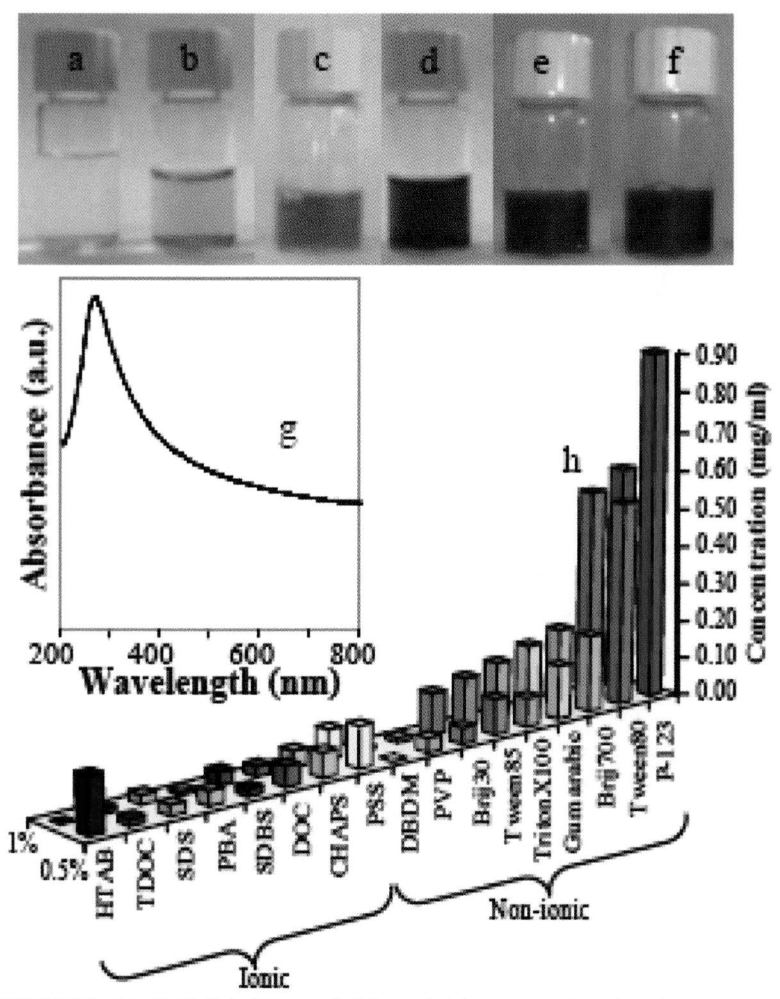

FIGURE 5.5. *(a)–(f) Digital pictures of vials containing only water (a) and aqueous graphene dispersions stabilized by 0.5 % wt/vol of TDOC (b), PBA (c), CHAPS (d), Brij 700 (e) and P-123 (f). (g) Typical UV-vis absorption spectrum of P-123-stabilized graphene dispersion. (h) Concentration of graphene in aqueous dispersions achieved by the use of different surfactants, as estimated from UV-vis absorption measurements. Two surfactant concentrations are shown: 0.5 and 1.0 % wt/vol. (Adapted with permission from [5]).*

surfactants typically lie in the ~0.01–0.10 mg mL^{-1} range, whereas non-ionic surfactants yield concentrations of the order of ~0.10 mg mL^{-1}. Particularly noteworthy is the performance of Tween 80 and P-123, which achieve concentrations in the range of 0.5 to 1.0 mg mL^{-1}.

5.5. APPLICATIONS OF GRAPHENE COLLOIDS

The previous sections demonstrated the variety of techniques that are available to form stable colloidal suspensions of graphene oxide, functionalized graphene oxide, reduced graphene oxide, and graphene. In this section, I will focus on two very different and important applications of such colloids. The first is to spray coat graphene onto surfaces to control the wettability of the surface, and the second is graphene colloids as novel cutting fluids for micro-machining applications.

5.5.1. Graphene Coatings with Controllable Wetting Properties

Graphene colloids can be spray coated onto substrates or simply deposited drop-wise on the surface. When the solvent evaporates, what is left is an interconnected network of graphene sheets that is deposited on the surface [6]. Such graphene coatings are highly effective at controlling the wettability of the substrate onto which they are deposited. Controlling the wettability of a solid surface has many practical applications. In particular, extremes in wetting behavior are highly desired. For example, super-hydrophobic materials with water contact angle above 150° are the key enablers for anti-sticking, anti-contamination, and self-cleaning technologies [7–12]. Similarly, super-hydrophilic materials with water contact angle below 10° have many important applications; for example, as a wicking material in heat-pipes and for enhanced boiling heat transfer [13–14].

In order to disperse the graphene sheets on a substrate, high power ultra-sonication of the graphene in a suitable solvent such as water or acetone is performed. The desired amount of graphene sheets is first weighed and dispersed in the solvent (ratio of ~100 mL of solvent to ~0.05 g of graphene) using an ultrasonic probe sonicator at high amplitude for ~1 hour. To deposit the graphene film, ~10 cc of the suspension is released on the substrate by a dropper in a clean room environment and allowed to dry for 12 hours under a fume hood for evaporation of the solvent. Subsequently, the surfaces are further cleaned and dried using nitrogen gas. Figure 5.6(a)–(b) show scanning electron microscopy

(SEM) images of graphene films deposited on an aluminum substrate. To determine the average roughness of the graphene films, a surface profilometer can be used; Figure 5.6(c) shows the measured average surface roughness parameter (R_a) for Aluminum (Al), Gold (Au) and Highly Ordered Pyrolitic Graphite (HOPG) substrates before and after deposition of graphene films with pure water as the solvent (Graphene-W film) and with pure acetone as the solvent (Graphene-A film). The

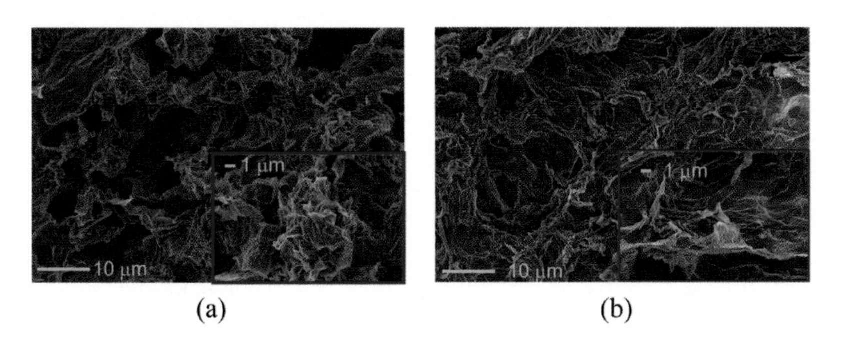

(a) (b)

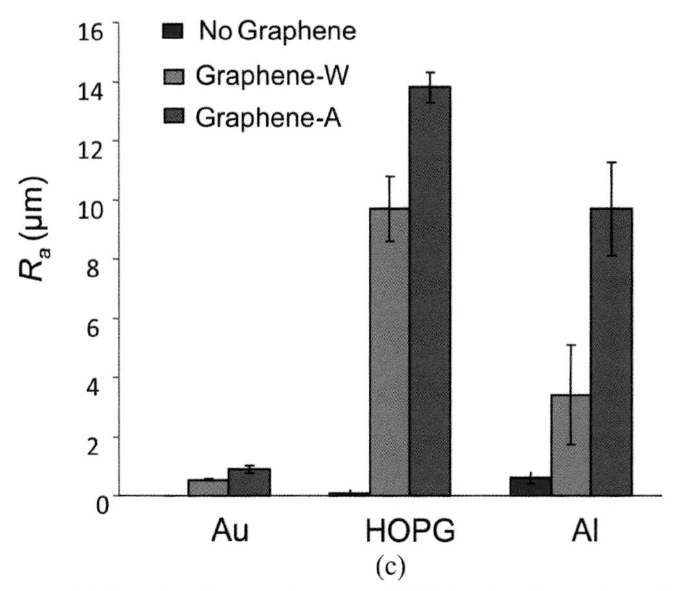

(c)

FIGURE 5.6. *(a) Scanning electron microscopy (SEM) imaging of a graphene film deposited on an Al substrate. Here, acetone is used as the solvent for graphene ultrasonication (Graphene-A Film). (b) Corresponding SEM image for graphene film deposited on Al substrate, but with water as the solvent used to ultrasonicate graphene (Graphene-W Film). (c) Average surface roughness parameter (R_a) for Al, Au, and HOPG substrates is compared before and after Graphene-W and Graphene-A film deposition. The graphene film enhances the surface roughness of the underlying substrate by 1-2 orders of magnitude. (Adapted with permission from [6]).*

graphene-coated samples show ~1–2 orders of magnitude increase in the average surface roughness as compared to the baseline substrates prior to graphene deposition. The average surface roughness of Graphene-A films was also observed to be larger than those of Graphene-W films. This is also confirmed by SEM imaging [insets in Figure 5.6(a)–(b)]. The wettability of the surface can be studied by performing static contact angle measurements by placing a droplet of de-ionized water on the surface of the various substrates. The contact angle can be measured using a digital goniometer. A water droplet is first generated using the automatic dispenser of the goniometer. The sessile droplet is formed by fixing the needle and approaching the substrate parallel to the needle direction with a very gentle feed rate of a few microns per minute. The axisymmetric drop shape analysis profile (ADSA-P) software [6] can be used for estimating the contact angle of water on the solid surface.

For the Au substrate [Figure 5.7(a)], the contact angle of the baseline Au (without graphene film coating) is ~76.3°. When graphene sonicated in water (Graphene-W) is deposited on the substrate, complete wetting is observed; i.e., the water droplet brought into contact with the solid spontaneously forms a film with a water contact angle of ~0°. The Graphene-W film, therefore, displays extreme super-hydrophilicity. The opposite effect was observed with graphene sheets sonicated in acetone (Graphene-A); the water contact angle of the Au substrate with Graphene-A coating is observed to lie in the super-hydrophobic range (~160°). Identical behavior is reproduced for HOPG [Figure 5.7(b)] and also for Al [Figure 5.7(c)] substrates. Films composed of Graphene-W imparted super-hydrophilicity to the Al and HOPG substrate, while Graphene-A based coatings impart super-hydrophobicity to the underlying substrate. By combining water and acetone in various proportions (e.g., 1:1 to 1:10) as the solvent, one can also tailor the water contact angle of the resulting graphene films over a wide range as illustrated in Figure 5.7(d).

To explain these results, let us consider the Wenzel model [15] for water droplets placed on rough surfaces. In the Wenzel model, the apparent contact angle on a rough surface in the homogeneous regime, is expressed as:

$$\cos\theta_W = r(\cos\theta) \tag{1}$$

where θ is the contact angle on the flat surface and r is the roughness ratio, defined as the ratio of the true area of the solid surface to its projection area. Since the roughness ratio (r) is always larger than 1,

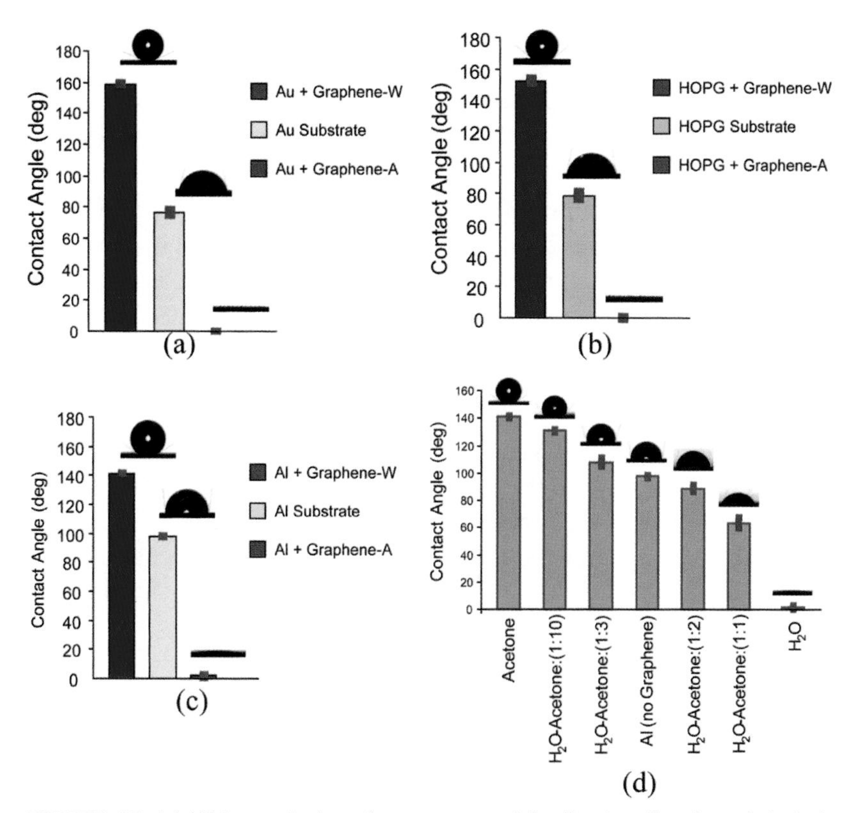

FIGURE 5.7. *(a) Water contact angle measurement for the baseline Au substrate is ~76.3°, which is increased to ~160° by Graphene-A deposition and decreased to ~0° by Graphene-W deposition. (b) For HOPG substrate, the baseline contact angle is ~81.5°, which is increased to ~151.4° by Graphene-A deposition and decreased to ~0° by Graphene-W deposition. (c) Corresponding results for an Al substrate showing variation in contact angle from ~97.5° for baseline Al to ~142.1° for Graphene-A film and ~1.6° for Graphene-W film. (d) Tailoring the contact angle of an Al substrate by ultrasonicating graphene sheets in a water/acetone mixture. Increasing the relative proportion of water in the mixture results in more hydrophilic behavior, while, conversely, higher acetone content yields more hydrophobic response. (Adapted with permission from [6]).*

from Equation (1) it is clear that θ_w will be higher than θ if the surface is originally hydrophobic (> 90°). Otherwise, θ_w is lower than θ. In other words, the roughness effect amplifies the inherent wettability of the substrate material. Therefore, if an individual Graphene-W sheet is hydrophilic, then a rough Graphene-W film [Figure 5.6(c)] is expected to be super-hydrophilic. Conversely, if an individual Graphene-A sheet is hydrophobic, then the Wenzel model predicts that a rough film [Figure 5.6(c)] composed of Graphene-A sheets would display super-hydrophobic response. Note that the as-produced graphene sonicated in water (Graphene-W) is expected to have carbonyl and carboxyl groups attached to the sheet edges, as well as residual epoxide and hydroxyl groups on the basal planes, which will make Graphene-W hydrophilic due to presence of carbon-oxygen bonds [16–17]. To understand why graphene sheets sonicated in acetone (Graphene-A) are hydrophobic, reflectance-mode Fourier Transform Infrared (FTIR) spectroscopy measurements were performed. The results [Figure 5.8(a)] indicate chemisorption of acetone on graphene for Graphene-A films, as evidenced by symmetric and asymmetric CH_3 stretching peaks [18] at ~2875 cm^{-1} and ~2964 cm^{-1} , respectively. Similarly, symmetric and asymmetric CH_2 stretching peaks [Figure 5.8(a)] at ~2852 cm^{-1} and ~2931 cm^{-1} are observed for the Graphene-A film. No such peaks were detected in the FTIR spectra of Graphene-W samples. Therefore, it appears that acetone is chemisorbed on the graphene surface [19–20] via C-O-C type bonds, in configurations where the terminal methyl (CH_3) groups point away from the nanotube surfaces. This view is also supported by ab initio calculations [21] based on density functional theory showing that acetone attaches preferentially at defects such as vacancies, via ~0.90–3.3 eV interactions resulting from C(acetone)-O-C(nanotube) bonds. For graphene sheets, defects (i.e., dangling bonds) are ubiquitous along the sheet edges. Moreover, exfoliation and reduction of graphite oxide to graphene leave the structure littered with defects such as 5- and 7-membered rings and carbon vacancies [22–24]. Furthermore, ultrasonication may also create new defects on the graphene basal planes, thereby facilitating the chemisorption of acetone. It is well established that methyl groups are hydrophobic in nature [25] and, hence, the attachment of terminal methyl groups to graphene will impart a net hydrophobic character to Graphene-A.

To summarize, graphene colloids can be used to deposit graphene coatings on substrates with controllable wetting properties. This provides a facile and effective means to modify the wettability of a variety of solid surfaces. Depending on whether water, acetone, or a combination of water or acetone is used as the solvent, the water contact angle of the surface can be tailored over a wide range from super-hydrophobic to super-hydrophilic behavior.

5.5.2. Graphene Cutting Fluids

Micro-machining operations such as micro-milling, micro-drilling, and micro-turning are currently being used to machine a wide-range

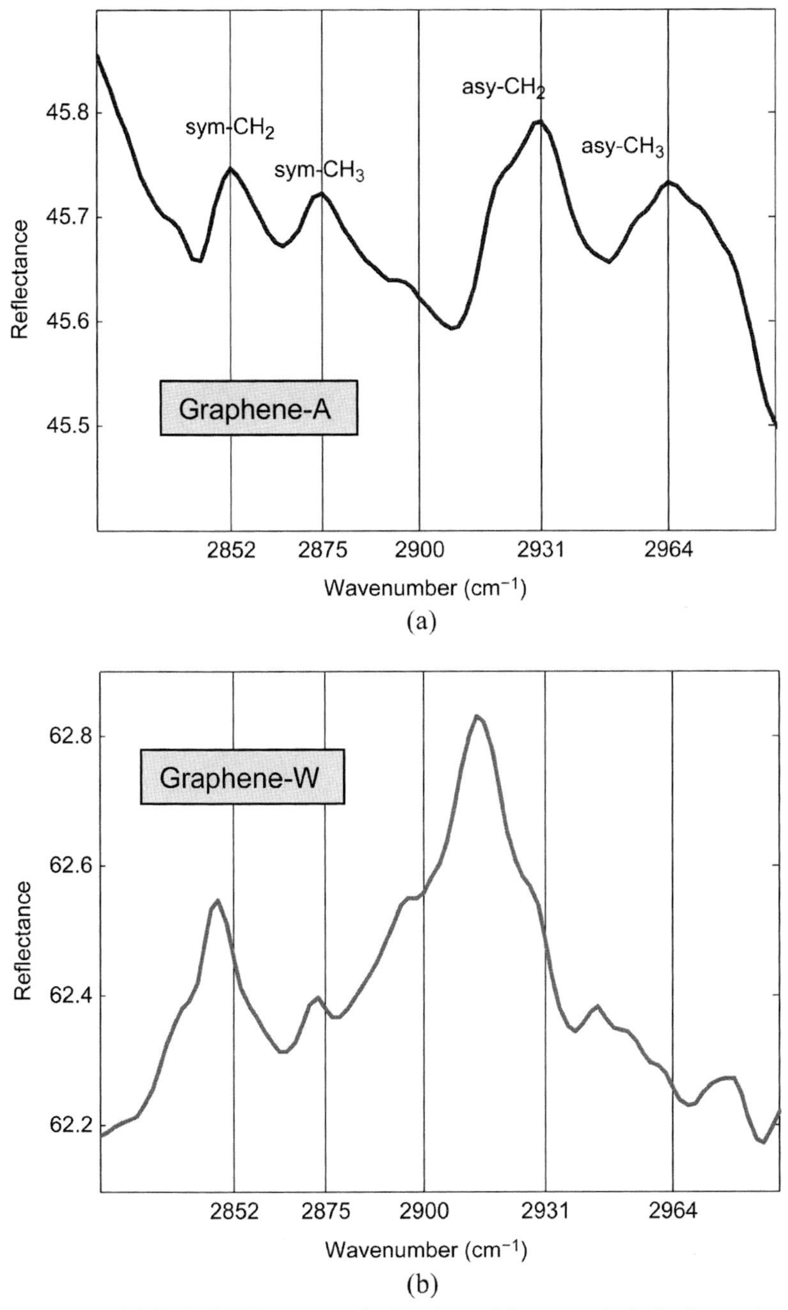

FIGURE 5.8. *(a) Typical FTIR spectra of a Graphene-A type sample, indicating peaks at ~2875 cm^{-1} and ~2964 cm^{-1} corresponding to CH_3 symmetric and asymmetric stretching modes, respectively. Similarly, peaks are also observed at ~2852 cm^{-1} and ~2931 cm^{-1} corresponding to the CH_2 symmetric and asymmetric stretching modes, respectively. (b) FTIR Spectra for a typical Graphene-W sample. No peaks in the spectra corresponding to CH_3 and CH_2 modes are detected for this case. (Adapted with permission from [6]).*

of precision micro-parts [26–28]. These manufacturing operations are high-strain-rate deformation processes characterized by extreme cutting temperatures and cutting forces that not only affect the life of the cutting tool, but also reduce the life-cycle of the micro-part by inducing thermal damage and residual stresses [29–30]. While novel techniques such as atomization-based delivery of micro-droplets into the cutting zone [31] have been shown to be successful for micro-machining processes, the overall performance of such a cutting fluid-delivery system is limited by the cooling and lubrication capacity of the cutting fluid used.

Historically, water-based semi-synthetic metal working fluids (MWFs) have been used as cutting fluids [32–34] for micro-machining applications. These MWFs consist of an oil-base that is dispersed in de-ionized water by using a surfactant. During the micro-machining operation, the lubrication is provided by the oil film, whereas the cooling is primarily provided by the evaporation of the water phase. Improving the lubrication efficiency of such an MWF invariably implies increasing the percentage of oil content in the MWF. However, this also increases the viscosity of the MWF, which drastically reduces the droplet delivery efficiency of the atomization system. Thus, there is a critical need to develop new MWF formulations that improve both the lubrication and cooling performance of MWFs without the expense of increasing its viscosity.

Nanofluids (engineered colloidal suspensions of nanofillers in a base fluid) may provide unique opportunities to enhance MWF performance [35]. The shape and size of the nanofiller additive is expected to significantly affect its suitability for micro-machining applications. The characteristic edge radius of cutting tools used in micro-machining processes is in the 1–2 μm range and the typical depth-of-cuts encountered are < 1 μm [31]. Ideally the thickness of the additives should be as small as possible so as to penetrate into the tool-workpiece/chip interface without affecting the dimensional accuracy of the part. Further, the lateral dimensions of the additive should be such that it effectively covers the 1–2 μm edge radius of the tool [31]. This combination of nano-scale thickness and micro-scale lateral dimensions are available in graphene and should enable graphene additives to penetrate the tool-workpiece interface while also covering the edge radius of the tool. This, in conjunction with their excellent thermal properties [36–38] (thermal conductivity > 1000 W/mK), make them well-suited for use as a performance-enhancing additive for MWFs in micro-machining processes.

Figure 5.9(a) depicts a schematic layout of the typical set-up used to perform micro-machining experiments. An atomization-based cutting fluid delivery system is used for transporting the MWF into the cutting zone. It consists of an ultrasonically vibrating piezoelectric transducer that is attached to a cutting fluid reservoir. The vibrations of the trans-

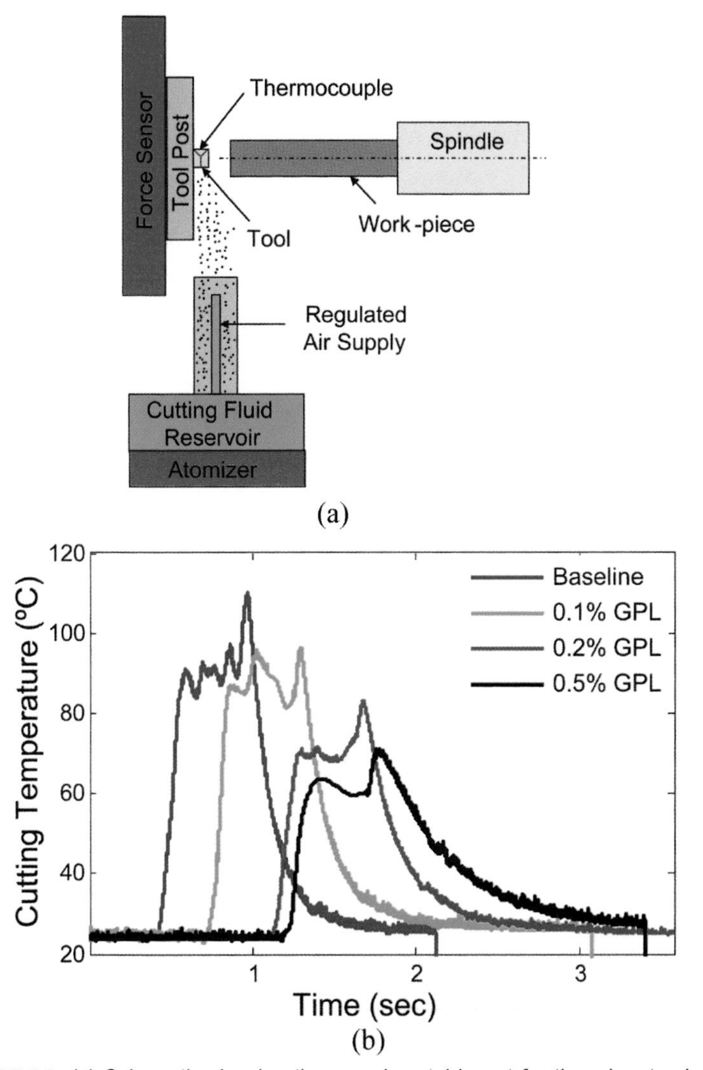

FIGURE 5.9. *(a) Schematic showing the experimental layout for the micro-turning operation. (b) Typical time trace for the cutting temperature on the tool during the cut for the baseline (pristine) MWF, as well as for various weight fraction of GPLs added to the MWF. The GPLs show a significant effect on suppressing the peak temperature of the tool during the cut. (Adapted with permission from [35]).*

ducer generate atomized droplets that are then carried to a delivery pipe. A co-axial air tube supplies the regulated air velocity needed to transport the atomized droplets into the cutting zone. The atomization-based cutting fluid system is typically attached to a three-axis micro-scale machine tool capable of performing micro-turning operations. This section presents proof-of-concept micro-turning experiments conducted on a 6 mm diameter rod of 1018 steel using a right-handed cubic-boron nitride micro-turning tool having an edge radius of ~2 μm. The cutting speed is ~250 m/min. The radial depth-of-cut, feed-per-revolution, and the total length of cut are maintained at ~40 μm, ~5 μm/revolution, and ~3 mm, respectively.

For the results presented here, semi-synthetic cutting fluid Castrol Clearedge 6519 at 12.5% dilution was used as the baseline cutting fluid. Graphene platelets (GPLs) obtained by thermal exfoliation and reduction of graphite oxide (Chapter 1) at loadings of 0.1%, 0.2%, and 0.5% by weight were added to the baseline cutting fluid to create three formulations of graphene-enhanced MWFs. Formulations containing 0.5% by weight of single-walled carbon nanotubes (SWNTs) and multi-walled carbon nanotubes (MWNTs) were also prepared to compare with the GPLs. The SWNTs (diameter ~2 nm, length ~10 μm) and MWNTs (diameter ~20 nm, length ~10 μm) were provided by Nanocyl. The graphene and nanotube fillers were dispersed in the cutting fluid by ultrasonication for ~10 min. Cutting temperatures and forces are the machinability metrics used to compare the performance of the various nanofluids. The cutting forces were measured using a Kistler 9018 triaxial load cell by sampling at a frequency of ~50 KHz. The cutting temperatures were measured using a type-J thermocouple (range 0–750°C) in conjunction with an Analog Devices 5B47 linearized thermocouple input module. The tip of the thermocouple was attached on the rake face of the tool at a distance of ~0.8 mm from the cutting edge.

Figure 5.9(b) shows time traces of the temperature measured by the thermocouple during the cut for the baseline MWF, as well as formulations of graphene-enhanced MWFs. Incorporation of GPLs in the MWF serves to significantly suppress the peak temperature of the tool during the cut. This effect is enhanced with increasing loading fraction of GPLs in the colloidal suspension. Another interesting observation is that the cutting temperature fluctuations with the GPL formulations are smaller than the baseline case. Since the thermocouple was mounted on the rake face of the tool, the temperature profile captures the heat generated at the tool-chip interface, which is primarily a function of the dynamic coefficient of friction at that interface. The fact that the GPL formulations result in lower temperature fluctuations points to a more uniform and lower coefficient of friction at the tool-chip interface as compared to the baseline. This suggests the ability of the GPLs to penetrate into the tool-chip interface in micro-machining processes.

Figure 5.10(a)–(b) depict the trends seen in the rise in cutting temperatures and the resultant cutting forces recorded during the course of the micro-turning experiments. Addition of graphene to the baseline MWF reduces both the average cutting temperatures and cutting forces during the cut. The addition of 0.1%, 0.2%, and 0.5% of graphene is seen to result in a ~5.6%, ~30.5%, and ~42% improvement, respectively, in the cooling performance of the baseline cutting fluid. The performance of 0.5% SWNT and 0.5% MWNT solutions is similar and they both lie in-between that of the 0.2% and the 0.5% graphene solutions. The cutting force shows an approximately linear decrease as the loading fraction of graphene is increased from 0% to 0.5% by weight. The lowest cutting force is observed for 0.5% GPLs, with its value being ~26% lower than that of the baseline cutting fluid. The 0.5% SWNT solution is seen to result in cutting forces comparable to the 0.1% graphene solution. The performance of 0.5% MWNT solution lies between that of 0.1% and 0.2% graphene solutions. The error bars in Figure 5.10 are indicative of fluctuations during the cut. The fact that the fluctuations are reducing with an increase in GPL content indicates a more stable machining operation. In summary, the results reveal that addition of graphene improves both the lubrication and the heat dissipation capabilities of the baseline cutting fluid. The improvements obtained in the cooling performance outweigh those seen in the lubrication performance. The performance of 0.5% GPL solution clearly out-performs that of the 0.5% SWNT and 0.5% MWNT formulations.

In order to understand the mechanisms that are underlying the trends seen in the machining responses, the contact angle (with the tool), the thermal conductivity, and the kinematic viscosity of the various MWF formulations were measured. As can be seen in Figure 5.11(a), the contact angle of the cutting fluid (measured against the tool surface) reduces continuously with an increase in graphene content. When compared against the baseline cutting fluid, an addition of 0.1%, 0.2%, and 0.5% graphene results in a reduction in contact angle by 43%, 52%, and 58%, respectively. The contact angles of solutions containing 0.5% SWNTs and 0.5% MWNTs are still higher than those of the 0.1% graphene solution. This indicates that, even with the addition of only 0.1% graphene,

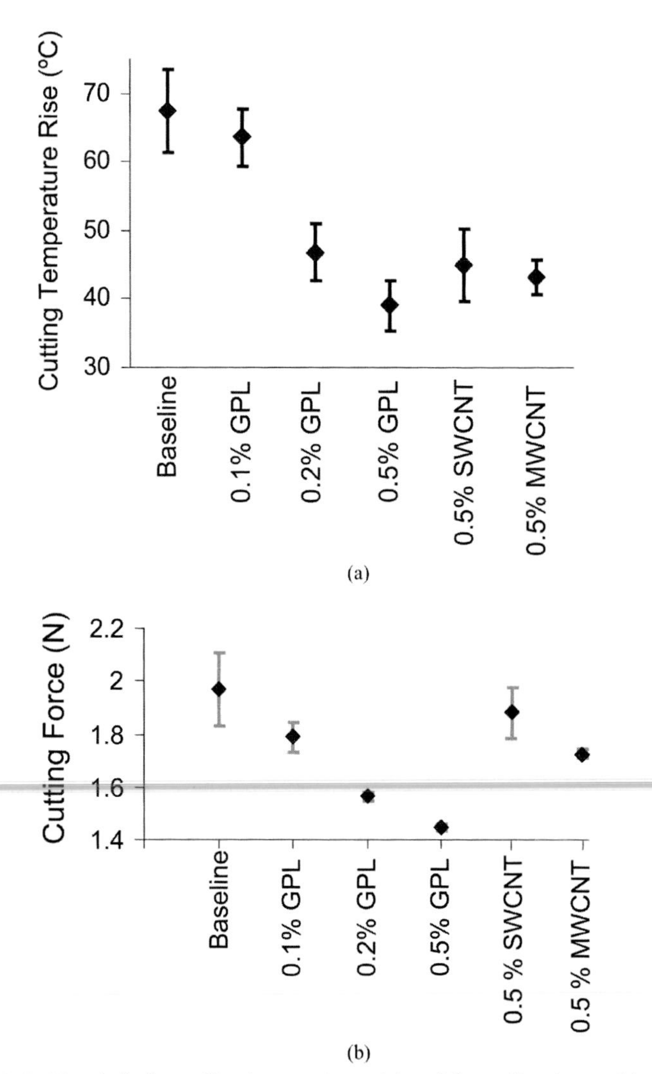

FIGURE 5.10. *Trends in the cutting temperatures (a) and the cutting forces (b) encountered for various nanoscale additives. Data is shown for the baseline MWF and for various nanofiller-enhanced MWFs. The lowest cutting temperatures and cutting forces were recorded for the ~0.5% weight GPL-MWF formulation. (Adapted with permission from [35]).*

the cutting fluid has a great propensity to wet the tool. Figure 5.11(b) depicts how the thermal conductivity of the cutting fluid increases with an increase in graphene content. When compared against the baseline cutting fluid, an addition of 0.1%, 0.2%, and 0.5% graphene results in an increase in thermal conductivity by 2%, 3% and 4%, respectively.

The thermal conductivity of the 0.5% SWNT solution is comparable to that of 0.5% graphene solution. The 0.5% MWNT solution has the second lowest thermal conductivity, with a value that is slightly above that of the baseline cutting fluid. The kinematic viscosity of the various MWF formulations can also be determined using a capillary viscometer. Figure 5.11(c) depicts that an addition of 0.1%, 0.2%, and 0.5% graphene results in an increase in kinematic viscosity of the MWF by 1%, 2%, and 3%, respectively. By contrast, the addition of 0.5% by weight of MWCTs and SWNTs results in a nearly 10% increase in the kinematic viscosity over that of the baseline cutting fluid.

The micro-turning results showed that both the cutting forces, as well as the cutting temperatures, reduce with an increase in the graphene content in the MWF. The data for the corresponding contact angles indicates enhanced wettability of the cutting fluid with the addition of graphene. In fact, the contact angle for even the 0.1% graphene formulation is lower than the corresponding values for the 0.5% SWNT and MWNT solutions. When wetted by the nanofluid, a coating of nanofillers is expected to form on the surface of the tool [39–40]. Residual oxygen groups on the GPLs (e.g., epoxide, carbonyl, carboxylic, or hydroxyl groups), in conjunction with the very high specific surface area of GPLs and affinity between the deposited GPLs and GPLs present in the bulk cutting fluid, will greatly enhance the wettability of the tool surface [25,41]. This extreme wettability of the graphene MWF formulations facilitates a more efficient entry of the GPLs into the tool-workpiece interface. The reduction in the cutting force implies that the graphene sheets appear to be successfully providing lubrication at the tool-workpiece interface. The fact that, for the case of graphene-based MWF formulations, there is only a modest 1–3% increase in kinematic viscosity that results in a 10–26% reduction in the magnitude of the cutting forces, indicates that lubrication is likely provided by the sliding of the graphene sheets and not by increased viscosity. It should be noted that, in the case of regular MWFs, since only the oil film provides the lubrication, an increase in viscosity of the MWF is absolutely necessary to ensure a reduction in cutting forces. This increased viscosity poses challenges to the effective delivery of such a cutting fluid using atomization-based techniques. The use of graphene platelet additives opens up the possibility of designing MWFs with improved lubrication efficiency without compromising heavily on the cutting fluid's viscosity.

The improvement seen in terms of the reduction in the cutting temperatures is very significant for the graphene-based MWFs. The low-

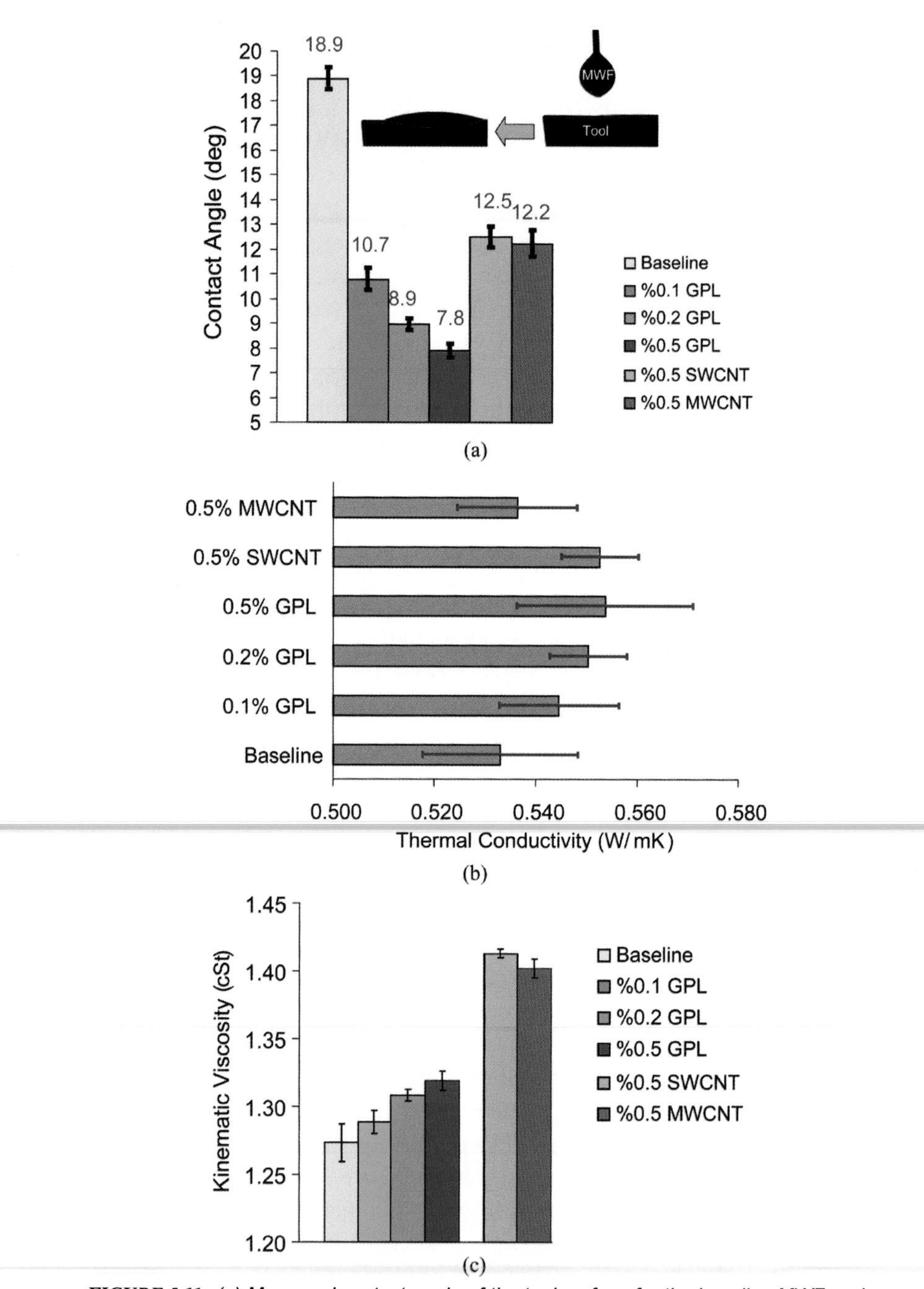

FIGURE 5.11. *(a) Measured contact angle of the tool surface for the baseline MWF and the nanofiller-modified MWFs. (b) Thermal conductivity measurements of the various MWFs used in the study. (c) Data for the kinematic viscosity (units of cSt: centi-Stokes) for the baseline and nanofiller-enhanced MWFs. The GPLs are superior to MWNTs and SWNTs at increasing the thermal conductivity of the MWF and the wettability of the tool surface with a minimal increase in the viscosity. (Adapted with permission from [35]).*

cutting temperatures for the graphene-based MWFs are likely to be the combined effect of both the improved lubrication (resulting in lower heat generation due to friction), as well as improved thermal conductivity of the MWFs with the addition of graphene (resulting in increased heat transfer from the cutting zone to the cutting fluid). Besides, the increased wettability of the graphene formulation implies that a greater surface area of the droplet is in contact with the workpiece. This implies more effective cooling provided by the evaporation of the water phase. Furthermore, the high thermal conductivity of graphene sheets is also expected to help conduct heat away from the cutting zone. The contact angle data reveals that the wettability of solutions containing SWNTs and MWNTs is significantly lower than that of the graphene-based solutions. Therefore, unlike the graphene platelets, the SWNTs and the MWNTs are not likely to penetrate into the tool-workpiece interface as effectively.

In summary, addition of GPLs to a semi-synthetic MWFs is observed to significantly improve its lubrication and cooling efficiency. This is attributed to the following factors: (1) Increased wettability of the graphene modified cutting fluid increases the solid-liquid interfacial contact area, enabling enhanced cooling by evaporation of the water phase. The improved wettability also facilitates the penetration of the graphene platelets into the tool-workpiece interface. (2) Effective lubrication provided by the sliding of the graphene sheets within the platelet. (3) Graphene sheets act as heat-sinks to shunt heat away from the cutting zone. GPL are also shown to be a superior additive for MWFs than either SWNTs or MWNTs. In the 0–0.5% GPL weight fraction range, there seems to be a nearly monotonic decrease in the cutting temperatures and the cutting forces, suggesting that further improvement may be possible by increasing the GPL loading beyond 0.5%. Such graphene-based colloids show potential to revolutionize the design of MWFs, especially for their use in micro-scale applications.

5.6. REFERENCES

1. Park, S.; Ruoff, R. S. Chemical methods for the production of graphenes. *Nat. Nanotechnol.* **2009**, *4*, 217–224.
2. Hazra, K. S.; Rafiee, J.; Rafiee, M. A.; Mathur, A.; Roy, S. S.; McLauhglin, J.; Koratkar, N; Misra, D. S. Thinning of multilayer graphene to monolayer graphene in a plasma environment. *Nanotechnology* **2011**, *22*, 025704.
3. Tang, X.-Z.; Li, W.; Yu, Z.-Z.; Rafiee, M. A.; Rafiee, J.; Yavari, F.;Koratkar, N. Enhanced thermal stability in graphene oxide covalently functionalized with 2-amino-4,6-didodecylamino-1,3,5-triazine. *Carbon* **2011**, *49*, 1258–1265.

4. Zhou, D.; Cheng, Q.-Y.; Han, B.-H. Solvothermal synthesis of homogeneous graphene dispersion with high concentration. *Carbon* **2011**, *49*, 3920–3927.
5. Guardia, L. ; Fernández-Merino, M. J.; Paredes, J. I. ; Solís-Fernández, P.; Villar-Rodil, S.; Martínez-Alonso, A.; Tascón, J. M. High-throughput production of pristine graphene in an aqueous dispersion assisted by non-ionic surfactants. *Carbon* **2011**, *49*, 1653–1662.
6. Rafiee, J.; Rafiee, M. A.; Yu, Z.- Z.; Koratkar, N. Superhydrophobic to Superhydrophilic Wetting Control in Graphene Films. *Adv. Mater.* **2010**, *22*, 2151–2154.
7. Lafuma, A.; Quere, D. Superhydrophobic states. *Nat. Mater.* **2003**, *2*, 457.
8. Furstner, R.; Barthlott, W. Wetting and self-cleaning properties of artificial superhydrophobic surfaces. *Langmuir* **2005**, *21*, 956.
9. Blossey, R. Self-cleaning surfaces—virtual realities. *Nat. Mater.* **2003**, *2*, 301.
10. Furstner, R.; Barthlott, W.; Neinhuis, C.; Miwa, M.; Nakajima, A.; Fujishima, A.; Hashimoto, K.; Watanabe, T. Effects of the surface roughness on sliding angles of water droplets on superhydrophobic surfaces. *Langmuir* **2000**, *16*, 5754.
11. Cheng, Y.; Rodak, D. E. Is the lotus leaf superhydrophobic? *Appl. Phys. Lett.* **2005**, *86*, 144101.
12. Feng, X. J.; Jiang, L. Design and creation of superwetting/antiwetting surfaces. *Adv. Mater.* **2006**, *18*, 3063.
13. Chen, L.; Wang, Z.; Wang, P. I.; Peles, Y.; Koratkar, N.; Peterson, G. P. Nanostructured copper interfaces for enhanced boiling. *Small* **2008**, *5*, 1403.
14. Chen, R.; Lu, M.; Srinivasan, V.; Wang, Z.; Cho, H. H.; Majumdar, A. Nanowires for enhanced boiling heat transfer. *Nano Lett.* **2009**, *9*, 548.
15. Wenzel, R. N. Resistance of solid surfaces to wetting by water. *Int. Eng. Chem.* **1936**, *28*, 988.
16. Stankovich, S.; Dikin, D. A.; Dommett, D.; Kohlhaas, K.; Zimney, E. J.; Stach, E. A.; Piner, R. D.; Nguyen, S. T.; Ruoff, R. S. Graphene based composites materials. *Nature* **2006**, *442*, 282.
17. Wang, Z.; Ci, L.; Nayak, S.; Ajayan, P. M.; Koratkar, N. Polarity-dependent electrochemically controlled transport of waterthrough Carbon Nanotube membranes. *Nano Lett.* **2007**, *7*, 697.
18. Ellis, A. V.; Vijayamohanan, K.; Goswami, R.; Chakrapani, N.; Ramanathan, L. S.; Ajayan, P. M.; Ramanath, G. Hydrophobic Anchoring of Monolayer-Protected Gold Nanoclusters to Carbon Nanotubes. *Nano Lett.* **2003**, *3*, 279.
19. Chakrapani, N.; Zhang, Y. M.; Nayak, S. K.; Moore, J. A.; Carroll, D. L.; Choi, Y. Y.; Ajayan, P. M. Chemisorption of acetone on carbon nanotubes. *J. Phys. Chem. B* **2003**, *107*, 9308.
20. Kudin, K. N.; Ozbas, B.; Schniepp, H. C.; Prud'homme, R. K.; Aksay, I. A.; Car, R. Raman spectra of graphite oxide and functionalized graphene sheets. *Nano Lett.* **2008**, *8*, 36.
21. Kyte, J. The basis of the hydrophobic effect. *Biophys. Chem.* **2003**, *100*, 193.
22. Schniepp, H. C.; Li, J. -L.; McAllister, M. J.; Sai, H.; Herrera-Alonso, M.; Adamson, D. H.; Prud'homme, R. K.; Car, R.; Saville, D. A.; Aksay, I. A. Functionalized single graphene sheets derived from splitting graphite oxide. *J. Phys. Chem. B* **2006**, *110*, 8535.
23. McAllister, M. J.; Li, J. -L.; Adamson, D. H.; Schniepp, H. C.; Abdala, A. A.; Liu, J.; Herrera-Alonso, M.; Milius, D. L.; Car, R.; Prud'homme, R. K.; Aksay, I. A. Single sheet functionalized graphene by oxidation and thermal expansion of graphite. *Chem. Mater.* **2007**, *19*, 4396.
24. Rao, C. N. R.; Sood, A. K.; Subrahmanyam, K. S.; Govindaraj, A. Graphene: the new two-dimensional nanomaterial. *Agewandte Chemie* **2009**, *48*, 7752–7777.
25. Park, S.; An, J. A.; Jung, I.; Piner, R. D.; An, S. J.; Li, X.; Velamakanni, A.; Ruoff, R. S. Colloidal suspensions of highly reduced graphene oxide in a wide variety of organic solvents. *Nano Lett.* **2009**, *9*, 1593.
26. Liu, X.; DeVor, R. E.; Kapoor, S. G.; Ehmann, K. F. The mechanics of machining at the micro-Scale: Assessment of the current state of the science. *J. Mfg. Sci. and Eng.* **2004**, *126*, 666.

27. Melkote, S. N.; Kai, L. Effect of plastic side flow on surface roughness in micro-turning process. *Int. J. Mach. Tool Manu.* **2009**, *46*, 1778.
28. Tansel, I. N.; Arkan, T. T.; Bao, W. Y.; Mahnedrakar, N.; Shisler, B.; Smith, D.; McCool, M. Tool wear estimation in micro-machining. I. Tool usage-cutting force relationship. *Int. J. Mach. Tool Manu.* **2000**, *40*, 599.
29. Torres, C. D.; Heaney, P. J.; Sumant, A. V.; Hamilton, M. A.; Carpick, E. W.; Pfefferkorn, F. E. Analyzing the performance of diamond-coated micro end mills. *Int. J. Machine Tool Manu.* **2009**, *49*, 599.
30. Wang, B.; Liang, Y. C.; Zhao, Y.; Dong, J. Measurement of the residual stress in the micro milled thin-walled structures. *J. Physics: Conf. Series* **2006**, *48*, 1127.
31. Ghai, I.; Wentz, J.; DeVor, R. E.; Kapoor, S. G.; Samuel, J. Droplet Behavior on a Rotating Surface for Atomization-based cutting fluid application in micro-machining. *J. Mfg. Sci. Eng.* **2010**, *132*, 0110171.
32. Jun, M. B. G.; Joshi, S. S.; DeVor, R. E.; Kapoor, S. G. An experimental evaluation of an atomization-based cutting fluid application system for micro-machining. *J. Mfg. Sci. Eng.* **2008**, *130*, 0311181.
33. Vieira, J. M.; Machado, A. R.; Ezugwu, E. O. Performance of cutting fluids during face milling of steels. *J. Mat. Process. Tech.* **2001**, *116*(2–3), 244.
34. Skerlos, S. J.; Hayes, K. F.; Clarens, A. F.; Zhao, F. Current advances in sustainable metalworking fluids research. *Int. J. Sust. Mfg.* **2008**, *1*, 180.
35. Samuel, J.; Rafiee, J.; Dhiman, P.; Yu, Z. -Z. and Koratkar, N. Graphene colloidal suspensions as high performance semi-synthetic metal-working fluids. *J. Phys. Chem. C* **2011**, *115*, 3410–3415.
36. Ghosh, S.; Bao, W.; Nika, D. L.; Subrina, S.; Pokatilov, E. P.; Lau, C. N.; Balandin, A. A. Dimensional crossover of termal transport in few-layer graphene. *Nat. Mater.* **2010**, *9*, 555–558.
37. Faugeras, C.; Faugeras, B.; Orlita, M.; Potemski, M.; Nair, R.R.; Geim, A.K. Thermal conductivity of graphene in corbino membrane geometry. *ACS Nano* **2010**, *4*, 1889–1892.
38. Balandin, A.A.; Ghosh, S.; Bao, W.; Calizo, I.; Teweldebrhan, D.; Miao, F.; Lau, C. N. Superior thermal conductivity of single-layer graphene. *Nano Lett. 2008*, *8*, 902–907.
39. Kim, S. J.; Bang, I. C.; Buongiorno, J., Hu, L. W. Effects of nanoparticle deposition on surface wettability influencing boiling heat transfer in nanofluids. *Appl. Phys. Lett.* **2006**, *89*, 153107.
40. Wasan. D. T.; Nikolov A.D. Spreading of nanofluids on solids. *Nature* **2003**, 423, 156–159.
41. Li, D.; Müller, M. B.; Gilje, S.; Kaner, R. B.; Wallace, G. G. Processable aqueous dispersions of graphene nanosheets. *Nat. Nanotech* **2008**, *3*, 101–105.

Index